Treasured Trees

Treasured Trees

Paintings by Masumi Yamanaka
Text by Christina Harrison & Martyn Rix

KEW PUBLISHING
ROYAL BOTANIC GARDENS, KEW

Contents

Foreword 7
TONY KIRKHAM

Artist's preface 10
MASUMI YAMANAKA

A history of Kew in trees 21
CHRISTINA HARRISON

Recommended reading 122
Artist and authors' biographies 125
Index 126

The trees 33
MARTYN RIX

Sweet chestnut 34
Cedar of Lebanon 38
Japanese pagoda tree 44
Maidenhair tree (ginkgo) 48
Black locust tree 52
Oriental plane 58
Lucombe oak 62
Tulip tree 64
Turner's oak 70
Corsican pine 74
Stone pine 76
Chestnut-leaved oak 82
Giant sequoia and coast redwood 86
Armand's pine (Chinese white pine) 90
Handkerchief tree (dove tree) 94
Indian horse chestnut 98
Bhutan pine 108
Nikko maple 110
Indian bean tree (southern catalpa) 112
Goat horn tree 114
Bogong gum 116
Sapphire dragon tree 118

Foreword

TONY KIRKHAM

I was inspired to write this foreword after leading a group of tree enthusiasts around the Arboretum at Kew, showing them many of the interesting and notable heritage and treasured trees and many young, newly planted specimens, which maybe in another hundred years might become heritage trees too. I often liken an arboretum to a book – an encyclopaedia of history, geography, science and English. Walking through an arboretum, looking at each tree, is like flicking through that book – each page is a tree and each tree has a story with elements of everything you would expect to find in that encyclopaedia.

An arboretum should also be like a family with great grandparents, grandparents, parents and children, all representing the different ages of tree planting. The Arboretum at Kew lost 50 years after the two world wars, when little, if any, tree planting was done, and to the expert eye this is noticeable in the age imbalance across the Gardens, but since the Great Storm in 1987 tree planting has been a priority at the forefront of all horticultural operations in Kew's Arboretum. Many of the gaps created by trees lost in the storm are now filled with new introductions or reintroductions from expeditions to various botanically-rich temperate parts of the world. During the night of 16 October 1987 Kew lost 700 specimen trees in the Arboretum and it took over three years to clear the casualties. Across the south-east of England we lost an incredible 15 million trees, which matched the loss of elms to Dutch elm disease over a 20 year period. It was a real wake-up call to everyone that we must respect our 'treescape' and give it the attention and respect it deserves, and since then many new tree planting schemes have been born.

I love tree planting and found it one of the most rewarding aspects of my job – leaving a living legacy for my successors and the next generation of dendrologists and visitors to Kew, establishing a tree that one day might just be a potential heritage tree in a World Heritage Site. I have had the privilege of selecting the tree species to plant, sourcing the seed to grow in the nursery, marking the planting position in the landscape in relation to the scientific, historic, educational or purely aesthetic layout of the Arboretum and watching it establish and grow into a mature denizen of the collection.

One heritage tree that was admired on our Arboretum tour was the Japanese pagoda tree, *Styphnolobium japonicum*, known as one of the 'Old Lions' of Kew, with a planting date of 1762. It has many historical stories that you can read about on page 44, but for me this is a 'character tree' – a tree with an unusual form or shape. Most importantly this specimen is a living library of arboricultural works: a valuable archived collection of every aspect of tree surgery that has been carried out on it over its 262 years life in the Arboretum, including cavity work, bracing, propping, root decompaction, feeding and even a fence to prevent people from climbing on it to have their photograph taken with a venerable old veteran. Directly opposite, on the other side of the road, there is a younger specimen about to take on the successional role of the species when the old veteran finally fades. I planted this tree in 1986; it was grown from a seed collected during a Kew plant expedition to South Korea in 1982. Trees grown from seed collected in the wild are important to the collection because they

are supported by field data and the herbarium specimens that were collected at the same time. With the latest arboricultural care and tree management this young tree may well join the list of notable heritage trees at Kew in the future.

Since arriving at Kew in 1978 as a diploma student, I have joined and led many expeditions around the world to collect tree seeds to increase the diversity of woody species at Kew. Along with tree planting it has to be one of the most rewarding experiences in my 43 years at the Gardens before retiring in 2021. Such expeditions have given me a passion for trees and left me with many memories of trees, forests and people from countries such as China, Chile, Japan, South Korea, the Russian Far East and Taiwan.

I took my guests around the outside of the Waterlily House to a tree that grows in the shadow of the magnificent chestnut-leaved oak, *Quercus castaneifolia* – an unrivalled Champion Tree of the Tree Register of the British Isles.

The tree I wanted to introduce them to was *Paulownia kawakamii*, the sapphire dragon tree from Taiwan (see page 118). Thirty years on, I can still remember collecting the seed of this species from a large tree in the forests of Taiwan, and this memory always pops into my mind whenever I visit this amazing specimen and others.

Theoretically it should not be growing as well as it is, as the climate from where it was collected is almost subtropical, but it has proved us all wrong and thrives in this sheltered spot. To work out what trees need in cultivation, we must observe how they grow in their natural habitat, copying nature in cultivation. This is such an important aspect of managing a large tree collection like the one at Kew and others around the world – getting the soil type, planting aspect and space right, and knowing when to intervene with some arboricultural help. Visiting woodland and forests around the world on expeditions has given myself and others that opportunity, and I am able to put into practice what I have observed and learnt, and to share that knowledge with others, both amateur and professional.

All the trees in the Kew Arboretum are national treasures. Now Masumi has captured these individuals on paper so well, we can all enjoy and remember them long after they have gone. Her artwork is an inspiration to me and to many others.

Tony Kirkham MBE, VMH
Retired Head of the Arboretum,
Gardens and Horticultural Services,
Royal Botanic Gardens, Kew

Artist's preface

MASUMI YAMANAKA

I first became fascinated with the world and work of the Royal Botanic Gardens, Kew in 2006 when I started to paint specimens from the collections. My interest in trees began when I created a series of paintings of the Indian horse chestnut tree, *Aesculus indica* 'Sydney Pearce' for which I was awarded the prestigious RHS Gold Medal for Botanical Art in 2010. Kew is home to over 12,000 trees, of which a small group are considered 'heritage' or historic trees, while over 300 are Champion Trees – the largest and finest of their kind in the country.

The seed had been sown and my interest in these remarkable trees began to grow. I started to look at the heritage trees in all their beauty and would spend my days visiting each specimen. It was then that I decided to record each individual tree as a permanent archive for Kew's art collections.

When people asked why I decided to undertake this project, I always answered, half-jokingly, 'because the trees asked me to'. Since completing this work, Britain has experienced several unusually strong storms, taking their toll on the trees at Kew, including individuals that I had painted or just started to paint. Some lost many of their large branches, the shape of others was altered. The devastation affected me deeply, and my only consolation was that I had captured these beautifully-shaped trees before they were damaged. My joke, that 'the trees had asked me to draw them', was prescient.

Some of the trees at Kew are over 300 years old. The Gardens are home to rare trees from all over the world, including specimens planted by Princess Augusta over 250 years ago, brought back by plant explorers from distant continents, and, increasingly, plants in danger of extinction in their native habitat. I feel privileged to have recorded a part of this living archive of arboreal history.

As I painted and closely studied my chosen trees, I became very aware of how they support the natural environment and are an intrinsic part of our planet's biodiversity. Not only do they delight the eyes with their beautiful shapes and multiple shades and tones, they are essential to our ecosystem – from the exchange of gases which support life on Earth, to providing habitats for fungi, insects, birds and other animals.

As a botanical artist, I see my role as being not only to record and illustrate the trees, but also to capture their beauty and power. My aspiration is that my paintings will inspire those who see them to share my experience and respond to these magnificent trees at Kew, and appreciate more deeply the fundamental importance of trees to our planet.

I would like to offer thanks to the following staff members at Kew Gardens: former head of Library and Archives, Chris Mills; former head of publishing, Gina Fullerlove; former galleries and exhibitions leader, Laura Giuffrida; former illustration curator, Marilyn Ward; and editor of *Curtis's Botanical Magazine*, Martyn Rix, who all made it possible for me to publish this book and hold the exhibition. I would also like to thank the Library and Archives staff and botanists at the Herbarium who provided me with support throughout this project.

I have the greatest respect for all the Kew Gardens' staff and volunteers who dedicate themselves to protecting plants and our planet Earth. I am honoured to be one of them.

Masumi Yamanaka

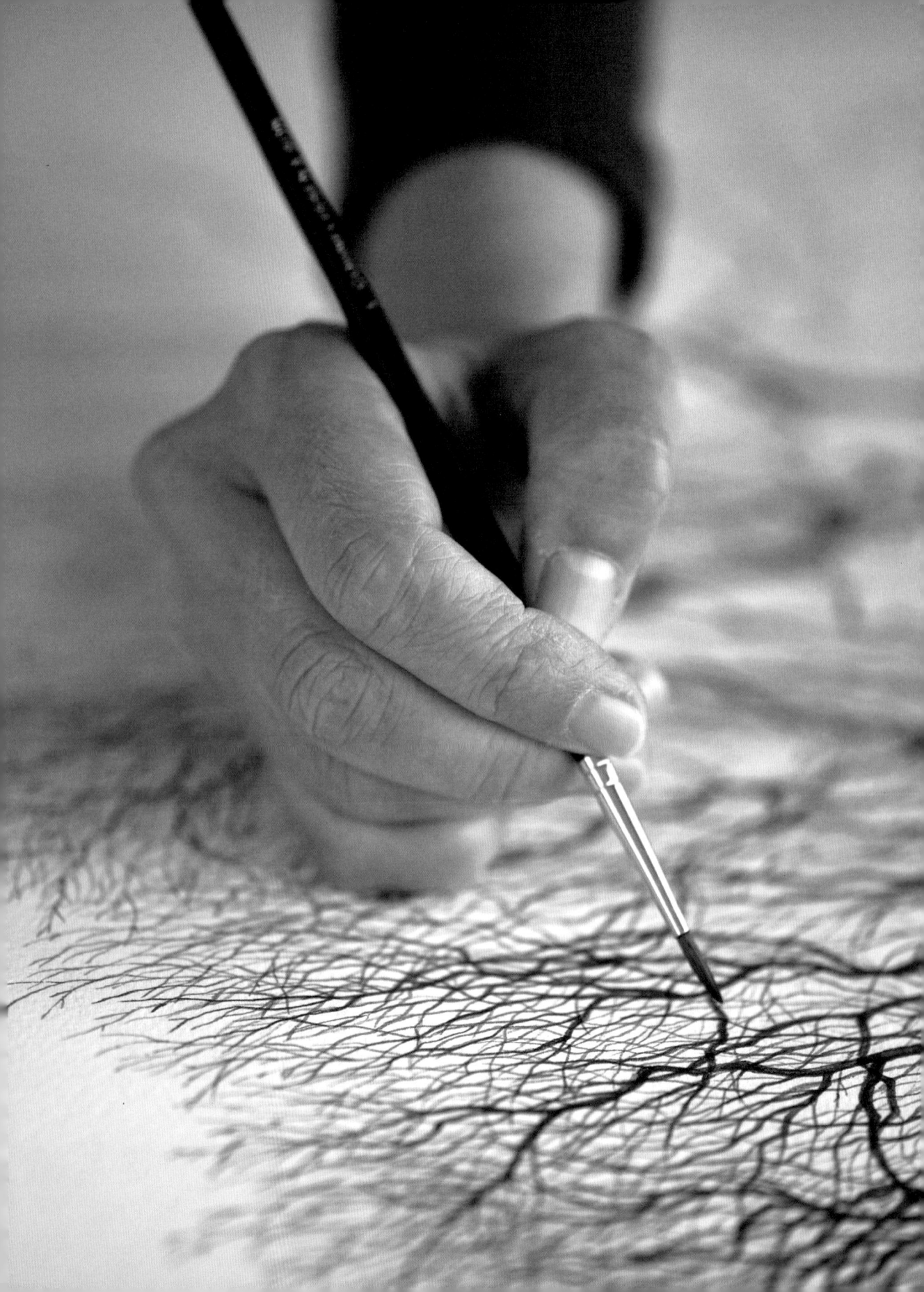

私がキュー王立植物園にて植物を描き始めたのは2006年のことである。世界最大の植物園のコレクションは「宝庫」としか言いようがなく、アーティストにとって毎日が感動の連続であった。

　私の樹木に対する興味はインドトチノキの一年をシリーズで描き上げた頃から始まった。(この絵のシリーズは2010年に英国王立園芸協会(RHS)のボタニカルアート部門にてゴールドメダルを受賞している。)この木はキューガーデンの1万2千本以上の木の中でも特別な歴史のある木、"ヘリテージツリー"に指定された数少ない木々のうちの1本であり、またヘリテージツリー以外にもチャンピオンツリーと呼ばれる大木が300本以上あることを知った。私は興味を持った古い木々を毎日訪れ、観察し、その　年を通しての美しい姿に魅せられた。そしてこの木々をキューガーデンのアーカイブとして一本一本描きとめておこうと決心した。

　「どうして古い木を描き始めたのか」と聞かれると、いつも「木たちに頼まれたから」と冗談半分で答えていた。ところがこのプロジェクトを始めてから、イギリスは珍しく強風を伴う嵐に何度か見舞われ、私が描き終えた、また描き始めたばかりの何本かの木は多くの太い枝を失い、姿を変えてしまった。嵐の翌日、無残に折れた枝の切り口を見ていると悲しくて涙が止まらず、はからずも美しい姿の間に描き残しておいてよかったと思った。

　「木たちに頼まれた」という冗談がまるで予知されたことのようになってしまった。

　キューガーデンの樹木は古いもので樹齢300年を超える。250年以上前、皇太子妃だったプリンセス・オーガスタが植えたもの、かつてのプラントハンターたちが遠く離れた大陸から持ち帰ったもの、自生地では絶滅の危機にさらされているもの、ここには世界中から集められた貴重な木々が揃っており、まさにキューガーデンの"生きているアーカイブ"である。この歴史的な木々の肖像画を描く機会に恵まれたことを私は名誉に思っている。

　また樹木を観察し調べていくうちに、私は樹木がこの地球上の自然環境に影響を与える大きな役割を担っていることを深く意識するようになった。

　樹木はただ美しい姿と緑で私たちの目を楽しませてくれているだけではない。地球上の生命にとって最も必要な酸素を作り出す環境システムエンジニアであり、また、キノコ、昆虫、鳥、小動物といった数多くの生きものたちの生息地を提供してくれている。この豊かな地球上の自然環境と生物多様性を支えているのは樹木であると言っても決して過言ではないと思う。

　植物学者でも環境学者でもない私にとって樹木の役割を科学的に説明するのは難しい。ボタニカルアーティストの私がなすべきことはただ樹木の姿を描きとめるだけではなく、その美しさとパワーを表現することだと信じている。私の絵を通して多くの人々がキューガーデンの壮大な木々との繋がりを感じ、そしてこの地球上における樹木の存在、重要性を深く認識してくださるきっかけとなることを心より願う。

　最後に、この出版・展示会を実現させてくれた元図書館長クリス・ミルズ、元出版部長ジーナ・フラーラブ、元展示部長ローラ・ギフィリダ、元イラストレーション部学芸員マリリン・ワード、ボタニカルマガジン編集長マーティン・リクス、そして常に私を支えてくれた図書館のスタッフ、標本室の植物学者たちに厚く感謝の意を述べたい。

　植物を守るため、地球を守るため活動を続けるキューガーデンのスタッフを心から尊敬すると共に、自分自身もその一員であることを誇りに思う。

山中麻須美

↘ Indian horse chestnut with oriental plane
Aesculus indica 'Sydney Pearce' with
Platanus orientalis
Watercolour on paper
32 × 51 cm (13 × 20 in) detail

↓ → Oak leaves and acorns from Kew
Quercus spp.
Watercolour on paper
40 × 50 cm (16 × 20 in)

↓ → Young leaves, from left: chestnut-leaved oak, maidenhair tree, Japanese pagoda tree, tulip tree, Indian horse chestnut, oriental plane
Quercus castaneifolia, Ginkgo biloba, Styphnolobium japonicum, Liriodendron tulipifera, Aesculus indica 'Sydney Pearce' and *Platanus orientalis*
Watercolour on paper
29 × 49 cm (11 × 19 in)

A history of Kew in trees

CHRISTINA HARRISON

←
Lucombe oak
Quercus × *hispanica* 'Lucombeana'
Watercolour on paper
29 × 38 cm (11 × 15 in) detail

→ A selection of the choicest trees were planted around the Temple of the Sun, near the Orangery at Kew in the 1760s, for Princess Augusta's new five-acre arboretum. A few trees dating from this time can still be seen including a black locust tree, a maidenhair tree and a Japanese pagoda tree.

Trees are vital – they not only sustain us, but also inspire us. Over time people have had a varying relationship with trees and perhaps we do not revere them now as much as we once did or in the way they deserve. They are not only useful, but also uniquely beautiful and each has its own story to tell. Within Kew's 132 hectares there is one of the world's best collections of hardy trees with a history that is both fascinating and revealing. They have the ability to inspire both the gardener and the artist in all of us, if we just look closely enough.

Nearly every tree at Kew has a record of where it came from, who collected the seed, who planted it and when, and how it has been cared for over the decades. Many have been collected from the wild and proudly represent their species in this wide-ranging collection; others have been in the Gardens since before the formal design of the site began in the early 1700s. Some are even named after people who have worked here (see p. 94). These trees are the backbone of the Gardens, and through their stories the history and changing purpose of Kew is revealed.

Much has been written about plants in garden history, about plant hunting, garden fashions, and indeed how Kew was transformed from two royal pleasure gardens into a botanic garden of world renown. All of these topics go hand in hand to explain why thousands of plants including many enormous tree species, from across the world, have ended up in a garden next to the Thames. This book is by no means an exhaustive account of any of these topics, but is instead an introduction to, and a pure celebration of, some of the most characterful and important trees at Kew. Through Masumi Yamanaka's stunning paintings you can discover and appreciate Kew's heritage and Champion Trees, which are some of the best examples of their kind and some of the first of their species to be planted in this country. They offer us an intriguing insight into the past, revealing key moments in Kew's timeline and the history of garden design. They are testaments to man's desire to collect, curate and conserve. In particular, they are the embodiment of how fashion and botany came together at a unique World Heritage Site.

To have these exceptional trees immortalised in such fine botanical artworks is a rare treat. Masumi has skilfully captured the essence of these individuals and, more than that, has captured a moment in time as they live, grow and change.

THE TREE COLLECTORS
The craze for tree collecting is older than you might first expect, and certainly older than the Gardens at Kew. Indeed, many have dated it back to ancient Egypt when Queen Hatshepsut sent out collectors to the land of Punt to gather all manner of exotic goods and plants including 31 'myrrh' or incense trees. In terms of the trees at Kew however we need only step back a few centuries.

A new passion for plants arose in the 16th century when botanic gardens began to be founded across Renaissance Europe, and the thirst for botanical knowledge grew. New gardening publications began to be written, releasing botany and herbalism from a cloud of superstition and legend that had existed since the time of the early Greek, Roman and Islamic scholars. They extolled the many virtues of planting solely for pleasure and beauty instead of simply for function.

In William Turner's *A New Herball* (1551–68) there are details of how trees had been introduced to English gardens before the middle of the 15th century purely for ornamental purposes. These included the oriental plane (*Platanus orientalis*), stone pine (*Pinus pinea*), 'spruce fir', Italian cypress

(*Cupressus sempervirens*), and walnut (*Juglans regia*). Along with many introduced flowers and herbs they added variety and a new exotic flavour to the English garden, and proclaimed the wealth, connections and education of the garden owner.

While the 16th and 17th centuries brought an influx of extraordinary bulbs and flowers to English gardens, many from Ottoman Turkey, the 17th and 18th centuries brought many new trees and shrubs from North America. These played a large part in the establishment of a new English gardening style. It was at this crucial time that the gardens at Kew began to be formed.

In the 1600s, John Tradescant became one of the first 'plant-hunters'. He began a phase of collecting that brought a whole new palette of colours and textures to the English garden for both visual delight and education. He travelled widely in Europe, North Africa and Russia collecting all manner of new plants, and also obtained species from North America. It was an exciting time and Renaissance gentlemen and

→ Kew's expanding tree collections were organised by W. A. Nesfield from 1845. His plans grouped trees and shrubs taxonomically among a series of aesthetically-pleasing vistas and walks – an arboretum within a landscape garden.

their gardeners had to become botanically adept. Those who could afford it employed experts to stock their gardens and new imports and rarities were highly sought after and prized. At the same time much effort was put into trying to determine the correct environment for these new plants, and climate and soil were of much debate.

The Tradescants' (father and son) Lambeth garden contained practically every plant then known and their lists and records reflect it as an ark of botanical knowledge. John the Younger made three trips to Virginia in North America and introduced trees such as the American plane (*Platanus occidentalis*), the tulip tree (*Liriodendron tulipifera*), the black locust tree (*Robinia pseudoacacia*) and the swamp cypress (*Taxodium distichum*).

Henry Compton (1623–1713), Lord Bishop of London, was one of the earliest English tree collectors and experimenters. His garden at Fulham Palace (which can probably be considered Britain's first arboretum) was only 32 acres but here he introduced a wide range of new species. These came mainly from North America, through his contacts in the Church, and included *Magnolia virginiana*, *Liquidambar styraciflua* and *Acer negundo*. He introduced such a considerable number of plants that he made many available to the Brompton Park Nursery owned by George London and Henry Wise. This 100-acre nursery became famous for its variety and supplied many of the great aristocratic estates.

Patrons of the Brompton Park Nursery included John Evelyn who brought all the knowledge he had gained from his tours of the gardens of Europe and England to a wide audience in his work *Sylva, or a Discourse on Forest Trees* (1664). This book contains a wealth of information on what trees were being planted at the time. He mentions oaks, ash, chestnut, walnut, birch and lime, as well as mulberry, planes, pines and larch, cedar of Lebanon (*Cedrus libani*), cork oaks, cypress, myrtle and strawberry trees. Evelyn advised the planting of avenues as well as orchards and made a plea that 'wild' forests should be protected and cherished.

A REVOLUTION IN THE GARDEN

Throughout the 18th century the knowledge and use of plants vastly increased. The direction of botanical research was influenced by the experiments of Fellows of the Royal Society and helped by the new naming system developed by Carl Linnaeus. Botany became an approved pursuit, not simply for its intellectual worthiness but because it was seen to have great importance for many lines of business: tree selection for the shipbuilding industry, the improvement of commercial fruits, medicines and other products, not to mention the money that could be made from new horticultural hybrids.

A new mindset now pushed the creation of tree collections as well as the development of botany itself. It was against this backdrop of a horticultural and botanical revolution that enlightened gentlemen gardeners began to design and furnish their estates with interesting new trees. Important tree collectors included Lord Petre at Thorndon Hall in Essex, the Duke of Argyll at Whitton and Charles Hamilton at Painshill, Surrey, who strived to create a landscape of 'living paintings', while other great gardens such as Stowe, Goodwood and Longleat were also leading the way in terms of design. The numbers of trees bought for these estates and the number of species they planted are quite staggering. All were added to a new kind of natural-looking landscape, as suggested by the works of Batty Langley, Joseph Addison, Stephen Switzer and Alexander Pope – the first to propose what would become known as the English Landscape Movement. Pope's friends, the designers Charles Bridgeman and William Kent,

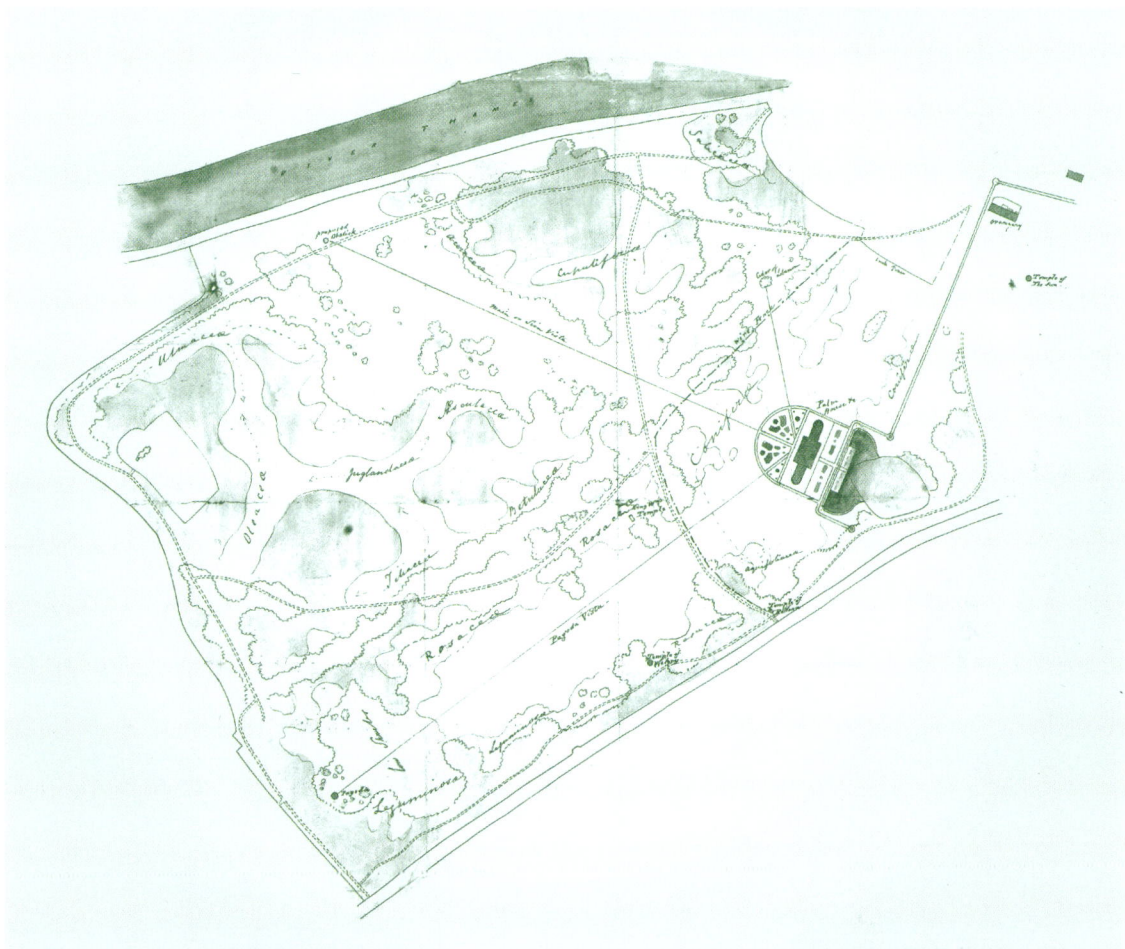

were to put these principles into practice on the estates of their clients to great acclaim. These pioneers were followed by Lancelot 'Capability' Brown (who also worked at Kew), and Humphrey Repton, who transformed many of the great estates of the country.

North American trees were in vogue, and were mainly supplied by a partnership between Peter Collinson, in England, and John Bartram, a botanist and farmer in Philadelphia, who sent over large boxes of seeds and small plants to be grown on in England. Bartram described himself as 'lit with a Botanik fire', and he later became known as America's first great botanist. Bartram also supplied Chelsea Physic Garden and Kew, as well as nurserymen such as James Gordon. A new breed of nurserymen had emerged, who were as eager for new plants and seeds as their clients, and their businesses developed apace. James Gordon's nursery at Mile End, Lee and Kennedy's nursery at Hammersmith and Christopher Gray's at Fulham were large suppliers of trees at this time. Many nurseries sprang up around London, particularly in Chelsea, in order to meet demand. Indeed it was the nurserymen, in the main, who were bringing new plants into the country and introducing them into the gardens of Britain. They sent out collectors with strict instructions to collect anything new and garden-worthy and built relationships with traders and sailors to bring new plants back from their travels overseas.

By 1742 Lord Petre had planted 60,000 trees at Thorndon, planting in subtle ways to entice the viewer out into the landscape, using height,

texture and colour. Ten thousand of these were from America including black walnuts, maples, red cedars, oaks and planes. Overall he had around 700 species on his estate. His collection was unrivalled and at the time of his death his nursery alone held 200,000 plants. Meanwhile, Horace Walpole labelled the Duke of Argyll a 'tree-monger' such was his desire for new trees, and he is recorded as having introduced 40 new species to Britain. His garden was known to have held over 350 species in total. After his death, in 1761, many of these admired trees were brought to Kew by his nephew Lord Bute.

It was this surge of tree collecting that vastly increased the choice of hardy trees for the designers of the English Landscape Movement, and changed the face of gardening in England. The 'discovery' of new species also encouraged the progress of botanical science and horticulture. It was an exciting time in the development of the English garden and almost everyone who could afford to embrace this new fashion eagerly did so.

It was at this time that Charles Bridgeman and William Kent redesigned the royal estate at Richmond for Queen Caroline. The Queen embraced the new English style and ordered new vistas, wildernesses, and follies to be built in her garden along the Thames, using the trees currently there from previous owners along with many new plantings. Bridgeman added many elms (*Ulmus*), beech (*Fagus sylvatica*), sweet chestnuts (*Castanea sativa*) and species of oak (*Quercus*) to create a sylvan landscape garden. He mixed woodlands with ornamental fields and used the ha-ha to extend the views.

Queen Caroline's son, Frederick, the Prince of Wales, took on the neighbouring Kew estate, which already had gardens of renown created by previous owners. Frederick greatly extended the gardens and planned many new plantings, follies and a lake. His vision for his garden went much further than

← Some of Kew's follies, such as the Pagoda, were set in aesthetically devised dense plantings of trees as recommended by the architect Sir William Chambers.

↓ Special boxes were invented to help plants survive the rigours of long sea voyages. This box was created in 1775 for transporting breadfruit and mangosteens 'or any other useful plants'.

his mother's, as he was extremely interested in the new botanical science of the day. Many of the trees he bought in were similar to those of other great landowners of the day – many from North America as well as species from Europe and Asia. He purchased hundreds from a nursery on Kew Green. However, his advisor, John Stuart, the third earl of Bute, was also able to bring in some rarities too. It seems to have been this friendship with Bute that sparked the desire to create a botanical collection at Kew rather than just a landscape garden. After Frederick's untimely death, his wife, Princess Augusta, completed his vision with the help of Lord Bute, whose botanical connections went far and wide. The first botanic garden was created in 1759, alongside the first small arboretum of special trees supplemented by Bute's uncle's collection from Whitton in 1762. By 1789 there were 630 species of tree listed in the arboretum, all arranged along wide gravel paths in an area of around five acres as part of a 9-acre botanic garden at the northern end of the Kew estate.

Some of the trees from this original arboretum are still thriving at Kew. One of the most visually interesting is the black locust tree on the Orangery Lawn, with its twisting, gnarly trunk held together with metal bands. The oldest trees from this time are known as the 'Old Lions' of Kew. They include the black locust tree, the large *Ginkgo biloba* nearby, the oriental plane near the Orangery and the elegantly reclining Japanese pagoda tree (*Styphnolobium japonicum*). The latter is thought to have been planted in 1762 and was one of the first five to be brought into the country by nurseryman James Gordon.

The fact that some of the trees to be planted at Kew in its first years as a botanic garden are still there as a reminder of the foundation of the institution is a wonderful testament to centuries of keen horticultural expertise. The next phase of tree planting came with an increase in the desire to make Kew a centre of botanical excellence. William Aiton had been hired in 1759 to curate and expand the collections and catalogue new species. He was one of the best horticulturists of his age and his skills helped to make Kew one of the finest collections in the country as many new choice trees were continually added to the gardens.

One of the greatest contributions to Kew in these early years came from Sir Joseph Banks, who, having returned from his voyage on the *Endeavour* with Captain James Cook, was keen to further the science of botany. Banks's passion for natural history and botany inspired King George III, who allowed Banks to send out plant collectors and set up a global botanical network to get hold of plants for Kew. In 1773 he took charge as Kew's first unofficial director. He began collecting species that would be useful for the nation, including growing plants for the colonies of the Empire. Banks was insistent that Kew should be the best botanic garden in the world, and have a representative collection from every

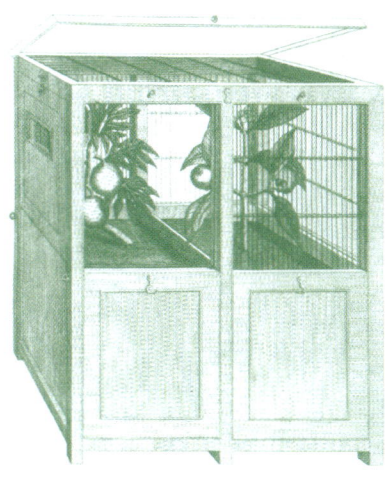

country. The garden became a dynamic place, and an active period of collecting, planting and reorganising began.

If Bute and Banks had ambitions for Kew, they were followed by two men whose visions were of an even greater magnitude. In 1841 the botanic garden was transferred to the government control of the Office of Woods and Forests, whilst the rest of the estate remained in royal hands as pleasure and hunting grounds. Sir William Jackson Hooker was appointed as the first official director. Hooker was an exceptionally qualified man for the job and his career had already been patronised by Banks. Under Hooker, and subsequently his son Joseph, the botanic garden took monumental leaps forward, expanding from 9 acres to covering almost the entire 300 acres of the estate, and became a leader in the botanical world once again. For the Hookers, botany was a beloved science and their life's work.

William Hooker was determined to renovate Kew. He planned museums, laboratories, a library, a larger herbarium, and, of course, the Palm House – which was to become the crowning glory of the Gardens and the focal point in a newly-planned arboretum.

By 1848, all of the Kew and Richmond estates came under the control of Hooker, encompassing around 250 acres. William Andrews Nesfield was brought in as a respected landscape designer, from 1844, to plan the main walks and avenues, and this later extended to a huge redevelopment of the site. He designed the Broad Walk, which was lined with tulip trees and deodars (*Cedrus deodara*), later replaced by Atlas cedars (*Cedrus atlantica*) as these proved better suited to the British climate. He also designed Syon Vista – originally an avenue of limes and deodars but eventually planted with holm oaks, and Pagoda Vista with matching pairs of ornamental trees. He mapped out a new arboretum according to the botanical families of the hardy trees that could be grown. Nesfield took into account aesthetic as well as botanical considerations; landscape design and botany were once again blended on the site. His plans involved felling many existing trees and it was at this time that most of the cedars around the Pagoda were taken out, leaving only four specimens of what was once a thick grove, and the Lucombe oak (*Quercus × hispanica* 'Lucombeana') (see page 62) was moved to make way for Syon Vista. Hooker bought in trees from nurseries across Europe. By 1850 Kew had set up new plant collecting arrangements with many other countries and institutions with help from the Admiralty.

The plantings of the mid 1800s were considerable in their number and variety, and many specimens are still thriving. At this time plant hunters were actively scouring the globe for botanical novelties. The early 19th century was an era of conifer collecting with many new species brought to Britain from western North America by the famous plant hunter David Douglas. Notable trees introduced in this era included conifers such as the sitka spruce (*Picea sitchensis*), the coast redwood (*Sequoia sempervirens*), and the giant sequoia (*Sequoiadendron giganteum*). Hooker developed a new pinetum next to the Palm House, where several large conifers can still be seen. By 1849, Hooker claimed there were 2,000 tree species and 1,000 varieties at Kew and he pronounced it a National Arboretum.

New avenues and a larger pinetum followed in the latter half of the 19th century, and thousands of new specimens were planted. One noteworthy individual planted at this time is the chestnut-leaved oak (*Quercus castaneifolia*) near the Palm House. This tree is now the largest and fastest-growing tree in the Gardens and is also a Champion Tree, being the largest of its kind in the country. Sadly, little is known of its origin or why

↓ The layout of Kew's botanic gardens has evolved over 250 years. This plan from 1847 shows the original botanic garden as well as a newly created pinetum, the Broad Walk and Palm House.

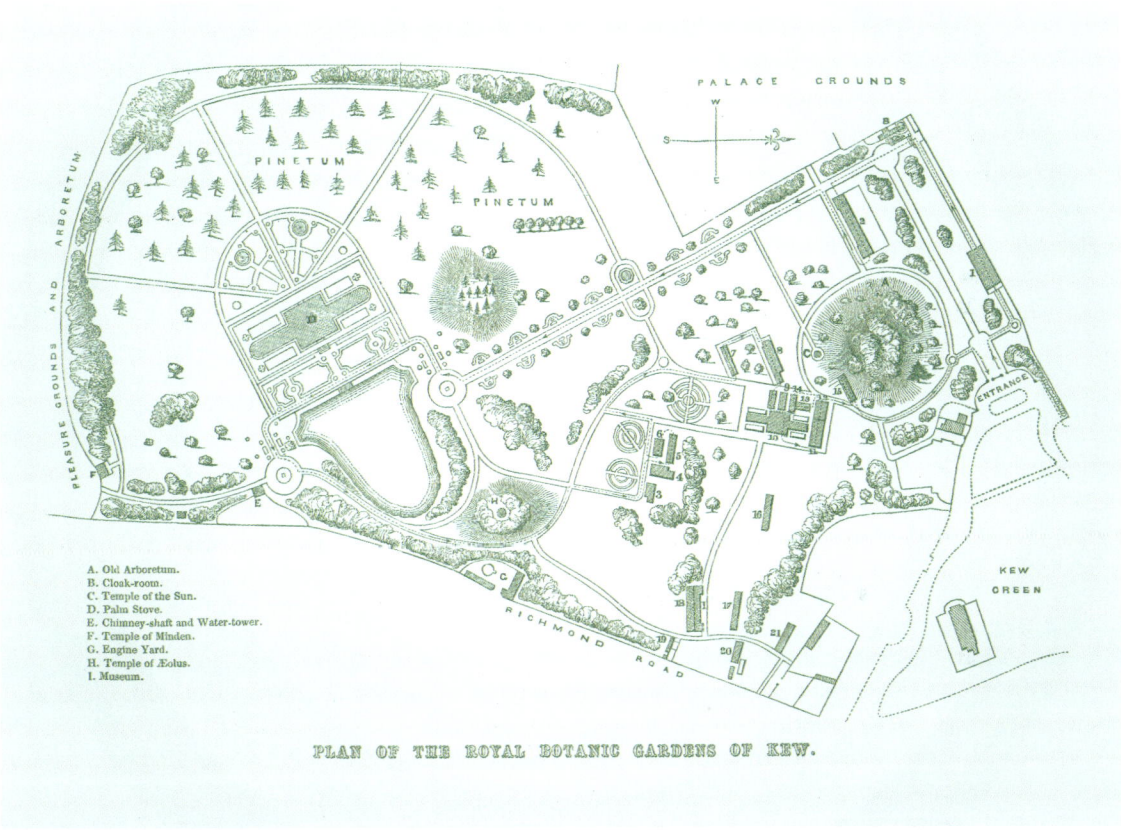

PLAN OF THE ROYAL BOTANIC GARDENS OF KEW.

A. Old Arboretum.
B. Cloak-room.
C. Temple of the Sun.
D. Palm Stove.
E. Chimney-shaft and Water-tower.
F. Temple of Minden.
G. Engine Yard.
H. Temple of Æolus.
I. Museum.

it was planted here, but we can admire its mature majesty today.

Planting continued into the 20th century and included new species from China, brought to Britain by famous plant hunters such as Ernest Henry Wilson. It is Wilson we must thank for bringing back seed of the beautiful handkerchief tree (*Davidia involucrata*), (see page 94) which no self-respecting arboretum would be without.

By the early 20th century, with this newly-designed and planned arboretum, its vistas, new glasshouses, and collections, the face of Kew had changed completely.

HISTORICAL LAYERS

Today, Kew has around 12,000 trees spread across the Gardens. Where you find a particular species can reveal much about the historical layers of the landscape. Kew's design is the result of many people's desires and planning over 265 years. The oldest trees on site date back to those royal estates at Richmond and Kew. The several ancient oaks

→ People flocked to see Kew's new Palm House and Waterlily House in the 1850s and enjoyed the wide variety of new plantings of species from around the world.

and sweet chestnuts, which appear in apparently random positions in the Gardens, are markers to where the royal estates had their 'wildernesses' and serpentine walks, carriageways and boundaries. You can see several ancient sweet chestnuts near the Mediterranean Garden and heading towards the Lake, while also at the Lake you'll find a substantial ancient English oak (*Quercus robur*), that now stands as a memorial tree to the lives lost during the Lockerbie disaster of 1988. Other ancient oaks, horse chestnuts and black locust trees can be found in the southern end of the gardens. There are other interesting specimens of a great age worth seeking out including a purple beech (*Fagus sylvatica* 'Purpurea') near the Ruined Arch, a black walnut (*Juglans nigra*) in the Woodland Garden, and a enormous plane tree (*Platanus × hispanica*) in the Rhododendron Dell, planted around 1773. To think that these trees might have been placed by some of the most influential designers of the day is a thrilling thought.

Other specimen trees on the northern lawns are remnants of surges of tree planting in the 1770s and the 1840s. Kew has hundreds of rare Champion Trees in the Gardens and the Tree Register states that Kew remains the richest garden for Champion Trees in the whole of Britain and Ireland, despite its sandy soil and the location near London affecting air quality. This is surely due to the long tradition of excellence in horticulture at the site and the immense care that has been taken over the planting and pruning of the specimens.

TREE COLLECTING TODAY

Trees are not just a part of our history. Tree collecting continues today – species new to science are still being found and new collecting expeditions take place every year to seek seed from species in the wild to increase the genetic diversity of Kew's collections and also to learn more about their native habitats. Sometimes such trips can collect seed from rapidly disappearing wild populations, and are of vast importance in terms of species conservation. There are thought to be around 60,000 tree species in the world. Kew's Millennium Seed Bank (MSB) currently holds seeds of more than 11,000 tree species. The MSB Partnership leads the Weston Global Tree Seed Bank Programme, which aims to collect seed from 3000 of the most threatened and useful tree species, and leads on the UK National Tree Seed Project, which is collecting seed from varied populations of 50 species of tree in the UK.

Kew's Arboretum, and the collections that now exist at Wakehurst, Kew's sister site in East Sussex, are of huge importance, particularly as they are attached to a scientific institution that researches and conserves wild tree species around the world. It is a living reference collection for all – not only for identification but also for genetic information and other research. Due to their age some individual specimens at Kew can hold invaluable genetic information, especially if the population they were collected from has since disappeared.

THE IMPORTANCE OF TREES

A complete hardy tree collection can help scientists in many ways. Much research has been done recently on choosing tree species that will thrive in our changing climate, and the importance of trees in the urban landscape – their ability to cool urban streets by as much as 6°C, and also their importance to our well-being, both physical and mental. Research into the benefits of living near a green space revealed a decrease in diseases, longer life spans and a reduced risk of obesity. A walk in the park raises spirits in a measurable way, lowers blood pressure, and can reduce the symptoms of attention deficit hyperactivity disorder in children. Interestingly it has also been found that the greater

diversity of species in that green space, the greater the benefits to mental health.

Other research focusing on carbon capture by trees has concluded that carbon sinks such as Kew's 12,000 trees take around eight tonnes of carbon out of the atmosphere each year. Trees are one of the most effective carbon sinks, and at a time when CO_2 levels in the atmosphere are at their highest, every contribution is important. Other studies have focused on the pollination services of trees, their value to biodiversity, and their value to the water cycle and the regulation of our climate. Tree canopies alone are said to hold 50 per cent of all life on our planet. In short, trees offer one of the most beneficial contributions to our environment.

The diversity of trees at Kew can prove invaluable to such research, as scientists work out which trees are best for what use. This is also the case for researchers looking at pests and diseases, and those looking at the value of trees for wildlife. As an arboretum the collection is also a priceless tool for education, both in the Gardens themselves and to a wider global community.

Trees have a wide-ranging importance today. We know they add form and structure to our gardens, parks and native landscape, but we also know them as valuable commodities for building, boundaries, tools, furniture, fibres, dyes, resins and fruits. They are also fundamental elements of forest ecologies, carbon sinks, wildlife hotels and a source of medicines. But to see them as just useful is often missing the point, for they are intrinsically beautiful and varied, and have evolved into an astonishing array of forms and types. This is what truly inspires collectors, gardeners and artists such as Masumi Yamanaka. Capturing the beauty and detail of individual specimens of the trees at Kew through the medium of botanical art is an extremely important task. The oldest trees are vulnerable, and some have changed or even died since these paintings were completed, due to weather and age. Masumi's artworks have captured a moment in time, a moment in the long history of Kew. They are a window into the stories of these beautiful trees, and the way Kew has grown and continues to develop.

The trees

MARTYN RIX

←
Indian horse chestnut
Aesculus indica 'Sydney Pearce'
Watercolour on paper
32 × 51 cm (13 × 20 in) detail

→ Sweet chestnut
Castanea sativa
Watercolour on paper
53 × 58 cm (21 × 23 in)

Sweet chestnut

Ancient specimens of the sweet or Spanish chestnut (*Castanea sativa*) are easily recognised by their gnarled trunks of furrowed and twisted bark. The specimen in this painting is believed to be from one of the oldest trees in Kew Gardens, and it may be a remnant of plantings on the royal estate in the early 18th century by the famous landscape gardener Charles Bridgeman. Sweet chestnuts were very popular at that time and planted in many of the gardens and parks where Bridgeman worked. A chestnut avenue was planted in Kew in 1880, and many large trees from this still survive in the Gardens.

The sweet chestnut is a native of southern Europe, where it forms forests in the mountains, usually above the evergreen Mediterranean zone and below the beech forests. Due to the fact that it was commonly planted by the Romans, its exact wild distribution is uncertain, but forests are now found from southern France and Spain to Turkey and North Africa. It is thought to have been introduced to England by the Romans, as the spiny outer coat of a fruit has been excavated by Hadrian's Wall, near Newcastle. The oldest chestnut tree still living is perhaps the Tortworth chestnut, growing near Stroud, which may be over 1,000 years old, but it is not the largest or the thickest, which is recorded as the seven sisters chestnut in Viceroy's Wood in Penshurst, Kent.

The Romans were avid planters of chestnuts on their estates in Italy. They were valued as a quick-growing crop for posts to support vines, or for charcoal production for sale to the cities, as they could be coppiced every five years. Their fruits were eaten, and still are, in a variety of ways – they are roasted whole, eaten as sweets when soaked in honey, or ground to make a flour that can be stored through the winter. The flowers of chestnuts are very rich in nectar and were an important source of honey; large forests of tall chestnuts and thousands of beehives can still be seen in the mountains of Euboea (Evia) in central Greece.

Chestnut coppice is a common feature of the woods in the Weald of Kent and Sussex, particularly on poor, sandy soils. The staves are split to make fencing, or kept whole for gateposts. Larger planks and beams were used in buildings as a substitute for oak for roof beams, as they are resistant to decay.

The leaves of sweet chestnut are simple, with stiff ribs and fine teeth along the margins; the flowers are small, the males made up of a mass of pale cream-coloured stamens, the females separate, showing the beginnings of the spiny husk. Usually the male flowers make up most of the flower spikes, with a few female flowers at the base, but Masumi has recorded that the tree she used for this painting has separate male and female flower spikes. The fruits are the familiar brown, shiny chestnuts with a tuft of small spines at their apex; they are enclosed by a spiny husk, velvety hairy inside. Generally only one out of three ovules in each husk are fertile; the other two are shrivelled and empty.

→ Cedar of Lebanon
Cedrus libani
Watercolour on paper
85 × 70 cm (33 × 28 in)

Cedar of Lebanon

The cedar of Lebanon (*Cedrus libani*) is possibly the most famous of all trees in the western world, praised for its strength, and for the great age it can reach. In the epic of Gilgamesh, in Sumerian, from before 2000 BCE, the cedar forests are noted for their great height and frightening density. Gilgamesh is said to have cut down the largest cedars to make a doorway for the temple and a raft to sail down the river Euphrates. Cedar logs were shipped to Egypt in around 2600 BCE for Thutmose III, and Hiram, King of Tyre, supplied cedar beams for King Solomon's temple. Pliny described a cedar trunk 40 m (131 ft) tall from Cyprus, used as a mast in around 300 BCE, and records trunks so thick that three men with outstretched arms could not encircle them.

Although there is much evidence that cedar forests were larger and more widespread in pre-Classical times, the Lebanon cedars still survive in a few protected sites in Lebanon and Syria, and there were cedar forests in the Anti-Lebanon Mountains too. Cedars are still common in Turkey, from the Amanus (or Nur) Mountains in the east, westwards along the southern edge of the Taurus Mountains to Muğla and Fethiye, where there are still fine forests on Baba Dağ, the ancient Mount Cragus, overlooking the sea opposite Rhodes. Young trees are upright, with a single trunk and pyramidal shape; they only develop the flat-topped shape characteristic of old trees when they reach around 200 years old. The oldest wild trees still living in Turkey are thought to be around 1,000 years old. There are also remnants of cedar forests in the Troodos Mountains in Cyprus; this is subspecies *brevifolia*, which has shorter, thicker needles than the mainland trees. The earliest cedars of Lebanon to be introduced to Britain and survive to maturity, were collected in the Lebanon in 1739 by the traveller the Reverend Richard Pococke, Bishop of Ossary and later of Meath; he described the great forest he saw there, and measured one of the largest at 7.3 m (24 ft) in circumference. The oldest specimens in England today were planted around 1760, some in the grounds at Highclere in Berkshire (where Pococke had been brought up), others at Goodwood in Sussex. There are other reports that a tree planted around 1646 by the Reverend Edward Pococke, rector at Childrey, near Wantage, still survives.

The Kew tree shown here is one of the older ones, probably planted in the late 18th century. It has now been felled, but was the much-admired, typical cedar of Lebanon shape, with a powerful upright trunk and horizontal, fan-like branches of shoots, producing a flat top. Trees take over 100 years to reach this shape, so we must be glad that so many were planted for future generations to enjoy.

The blue cedars, which originate from the Atlas Mountains in Morocco, are more recently introduced into gardens, and even the largest at Kew were probably planted in around 1880. They may yet form a more flat-topped shape, and apart from their leaf colour, be almost indistinguishable from the trees that originate in the Lebanon or Turkey.

↘ Cedar of Lebanon
Cedrus libani
Watercolour on paper
53 × 73 cm (21 × 29 in)

→ Japanese pagoda tree
Styphnolobium japonicum
Watercolour on paper
72 × 100 cm (28 × 39 in)

Japanese pagoda tree

After the unexpected death of her husband Frederick, Prince of Wales, in 1751, Princess Augusta continued his plans for the botanic gardens at Kew, and the area between the present Princess of Wales Conservatory and the Orangery was planted as an arboretum. This ancient specimen of the Japanese pagoda tree, *Styphnolobium japonicum*, (previously known as *Sophora japonica*) is one of Kew's original trees and was planted under the direction of William Aiton in 1762. It is known that it was one of five specimens imported in 1753 from China via France by James Gordon, an expert London nurseryman from Mile End, who introduced many new plants from North America as well as from China. It is said that the seeds were sent back from Peking along the Silk Road by Père d'Incarville in 1747.

Nowadays Kew's tree has two almost horizontal trunks, supported by steel props, and the base has a brick support that protects an aerial root. In an early photograph, taken in around 1910, there are four more upright trunks, but it is only these horizontal ones that have survived.

Because of its religious significance it was planted near temples in China and Japan, and this led to its common name, the Japanese pagoda tree. This tree is full of significance in China, and was often planted by the graves of scholars or senior mandarins, hence another common name, the scholar's tree. It is, however, wild in China and Korea, growing in dry valleys in the mountains, and in the Sichuan plain. There are large specimens in the old city of Beijing and around Chengdu, reaching a height of 20 m (66 ft), and a girth of 2–3 m (7–10 ft). It is only known in cultivation in Japan, and Linnaeus, who named it *Sophora japonica* in 1767, knew it only from a specimen he obtained from Dr Christiaan Kleinhoff, director of the Botanic Garden in Batavia (now Bogor).

It is widely used in Chinese traditional medicine; the fruit is an anthelmintic – used particularly (as E. H. Wilson noted) for treating parasitic worms in animals, and the flowers are used to kill other parasites; infusions of flowers and leaves can control insects. Extracts of the flowers are also reported to be effective in the treatment of piles, and the flowers can be used to produce a yellow dye, for silk and cotton. The timber is very hard and white, often used for pillars and door frames in Japanese temples.

This is a very hardy tree, and has spread in cultivation across Asia; for instance it was planted in the courtyards of madrasahs (centres of education), as it is in the Ulugh Beg Madrasah in Samarkand, Uzbekistan. It also grows well in North America, where it is often planted as a street tree. It grows quite fast, and produces a delicate shade, with many branches forming a rounded head. The flowers appear in late summer or early autumn, usually in August or September, and the branching sprays of pea-like flowers can be 30 cm (1 ft) long, covering the whole tree with a veil of creamy-white.

→ Maidenhair tree
Ginkgo biloba
Watercolour on paper
96 × 70 cm (38 × 28 in)

Maidenhair tree (ginkgo)

The maidenhair tree (*Ginkgo biloba*) is a unique survivor of earlier millennia. These elegant trees were widespread around 190 million years ago, in the early Jurassic, when dinosaurs roamed the Earth. Kew's tree is one of the oldest in Britain and is now a tall multi-stemmed specimen growing by the path at the edge of the old arboretum, not far from Elizabeth Gate on Kew Green. Because it came from China, early gardeners thought it was tender, and planted it against the wall of the Great Stove – an early type of hothouse built to house exotic plants in the royal garden in 1761. This was demolished in 1861 and the freestanding tree has grown well ever since. It was imported by James Gordon of Mile End Nurseries in 1754, and planted out around 1762.

This old tree is a male and produces yellow cone-like male flowers just as the leaves open. The mature leaves are fan-shaped and two-lobed, turning a beautiful bright yellow in autumn. This species is most often known by its Latin name *Ginkgo*, but its common name of maidenhair tree alludes to the shape of the leaves, which resemble those of the maidenhair fern. As the sun rises after an early autumnal frost, every leaf will fall in a few hours. The female trees, of which there is a young specimen at Kew near Kew Palace, produce yellow fruit like large cherries or small apricots; they have a large edible seed, surrounded by a smooth shell and thin rancid-smelling yellow pulp. The seeds are eaten in China and Japan, baked as a delicacy and eaten medicinally, and an infusion of the leaves is supposed to benefit the circulation and slow down memory loss and other ailments caused by poor circulation.

Wild ginkgo trees are almost unknown. They have been commonly planted near temples in China and Japan for many years and most of the large specimens are associated with ancient buildings. A famous specimen, at Li Jiawan in Guizhou, is nearly 30 m (98 ft) tall and just under 6 m (20 ft) across near the ground; it is hollow, now split into five main parts, and has four subsidiary trunks around 20 m (66 ft) tall. It is thought to be the oldest of all living ginkgo trees. The nearby temple has long since gone, so it stands alone in a deep, rural valley, surrounded by tobacco fields.

The oldest specimens in Japan are thought to date from around 1600, though the trees were certainly brought from China earlier than that, probably before 1450. There are legends that some large trees are even older, but they are hard to verify.

There have been many attempts by Chinese scientists to identify natural stands of *Ginkgo*, and they now suppose that populations on Jinfo Shan, near Chongquin, and on Tianmu west of Hangzhou near Shanghai, may be survivors from before the last glaciation. Ginfoshan, in particular, is known for other ancient trees, such as the conifers *Cunninghamia lanceolata* and *Cathaya argyrophylla*.

The specimen at Kew, known as one of the Gardens' 'Old Lions', is still thriving, and can be expected to survive another hundred years or more.

→ Black locust tree
Robinia pseudoacacia
Watercolour on paper
95 × 62 cm (37 × 24 in)

Black locust tree

This ancient specimen of the black locust tree (*Robinia pseudoacacia*) can be found near the Japanese pagoda tree and maidenhair tree (see pages 44 and 48), all part of the planting in Kew's original small botanic garden. This tree has several common names, and is usually called the black locust tree in America, perhaps because of its dark and gnarled trunk or even because of its dark calyx; its edible seeds reminded the early settlers of the locust tree or carob (*Ceratonia siliqua*), eaten by John the Baptist and now used as an alternative to chocolate.

Robinia is found wild in eastern North America, from Pennsylvania south to Georgia and west to Iowa and Oklahoma, and is now commonly naturalized in southern and eastern Europe. Because the plant has great clumps of nitrogen-fixing nodules on the roots, it can thrive in poor sandy soils. It also produces suckers as well as ample seeds, and can become a nuisance. Old specimens, which may reach 25 m (82 ft) in height, are very handsome and free flowering, with hanging bunches of white, sweet-scented flowers, much loved by honeybees. In France and Italy the famous acacia honey, a very clear and scented variety, is made from *Robinia* plantations.

Robinia was named by Carl Linnaeus after Jean Robin, who, with his son Vespasien, was gardener and herbalist to the Kings of France. The plant hunter John Tradescant was growing *Robinia* in England in 1634, and by 1640 it was noted as a fine tree in the garden of his son, John Tradescant the younger. The Robin family and the Tradescants were in regular communication, so it is not clear from where exactly the seed originated. Early botanists called this the acacia, or the locust tree of Virginia and hoped it would become an important timber tree in Europe.

This old *Robinia* was planted in 1762 and was one of the first trees in the new arboretum for Princess Augusta. It is one of the choice trees that came from the estate of the Duke of Argyll in Whitton, a famous tree collector of his day. It has been in a state of decline for many years now, but recently it has been rejuvenated by careful cultivation, and is now growing strongly. In its prime it would have been much taller than it is today, but the branches are vulnerable to wind damage, and the old top fell long ago. Nonetheless the tree survived a great storm of 28 March 1916, when a huge cedar of Lebanon nearby fell and demolished the William Chambers's Temple of the Sun, which was built at the same time as the *Robinia* was planted.

↗ Black locust tree, flower
Robinia pseudoacacia
Watercolour on paper
95 × 62 cm (37 × 24 in)

→ Oriental plane, with a ring-necked parakeet
(*Psittacula krameri*), now a common bird at Kew
Platanus orientalis
Watercolour on paper
50 × 70 cm (20 × 28 in)

Oriental plane

The towering oriental plane (*Platanus orientalis*) at Kew stands alone on the lawn in front of Kew Palace; it dates from the early days of Princess Augusta's arboretum and was probably planted in 1762. It is possibly a transplant from the Duke of Argyll's estate at Whitton near Hounslow. When first planted its hardiness was not known, and as it came from the Mediterranean area, it was planted against the east wall of the former White House, now demolished, whose outline is marked by stones in the lawn. It is said that three trees were planted in a row, but only this one has survived.

The oriental plane grows wild all around the eastern Mediterranean, across Turkey, and as far east as Iran. It always grows along stream-beds or springs, because the seeds need running water to germinate, and this habitat is often shared with the shrubby pink oleander (*Nerium oleander*). In spring the bright green young leaves contrast with the surrounding dark evergreens on the drier slopes. Like oleander, the trees grow happily away from water once they are established and are often planted in the middle of village squares.

Big old planes are among the legendary trees of the classical world. Hippocrates is said to have taught under a large plane tree on the eastern Aegean island of Cos in the sixth century BCE. Greek historian Herodotus relates how the Persian king Xerxes encountered a magnificent plane as he invaded Asia Minor and crossed the Meander river near Kallatebos. Impressed by its beauty, he ordered it to be guarded and presented it with golden ornaments. Cimon planted plane trees to give shade around the Parthenon in Athens in the fifth century BCE, and converted the Academy from a treeless square to a grove, laid out with shaded walks. Adolescent Spartan boys wrestled in an island grove, called the Planes, and Aristotle's plane tree in the Lyceum had roots that stretched 50 feet (15 m) along the ground. Newly civilized Romans learnt of the plane tree from Greek literature, and planted them in Rome and on their estates. Traditional Romans planted elms to support their vines, and Horace complains that the →

→ fashionable plane was driving out the elm. Pliny the Elder was even more 'anti-plane', most especially when the trees were pruned or trained to stay dwarf. Pliny's *Natural History* contains a list of famous plane trees and he recalls how he held a dinner party for 18 members of his staff in the hollow of a plane tree in Lycia.

Plane trees, known in Persian as the chenar (or chinar) tree, are typical of Persian gardens, and are commonly seen in Indian paintings. They were planted as far east as Kashmir, where they still line the watercourses of the Mughal gardens by Dal Lake.

Big, old planes are still found near many classical sites, notably along the stream at Pinara of ancient Lycia (now in Turkey), but few have survived to rival those planted in England. One of the oldest and biggest oriental planes is said to be the tree at Hawstead Place, near Bury St Edmunds, possibly planted in 1578. The tallest, a thinner tree which has reached 30 m (98 ft), is found in Petersham Lodge woodlands, not far from Kew, and has the benefit of groundwater from the nearby River Thames. The famous oriental plane at Christ Church in Oxford, brought from the Levant by Edward Pococke, professor of Arabic, and planted in 1636, is wider but shorter.

These oriental planes are smaller than the hybrid London planes (*Platanus* × *hispanica*), which are faster-growing; the oldest, and one of the largest, is near Ely Cathedral, and was first recorded in 1674. The huge London plane that towers over the Rhododendron Dell at Kew was planted in the 1770s, while other fine specimens can be seen near the Palm House Pond and the Ice House. One of the finest London planes in the country can be seen on the riverside not far from Kew, near Richmond Bridge.

It is not always easy to distinguish the two species of plane tree. The London plane tends to have more pale, flakey bark, a character of its American parent. The oriental plane tends to have more deeply-lobed leaves, most clearly seen in some forms from Crete, Cyprus and southern Turkey.

→ Lucombe oak
Quercus × *hispanica* 'Lucombeana'
Watercolour on paper
29 × 38 cm (11 × 15 in)

Lucombe oak

Old specimens of the Lucombe oak, *Quercus* × *hispanica* 'Lucombeana' (often known as *Q.* 'William Lucombe'), are among the finest evergreen oaks to be seen in England. Kew's gnarled old Lucombe oak is almost certainly a graft from the original cross, and is now a huge wide-spreading tree, to be found along the avenue of holm oaks (*Q. ilex*) that stretches from the Palm House to the River Thames. It was already a large tree in 1846, when it was moved 20 m (65 ft) to clear the view for the new Syon Vista, so it was probably planted in the late 18th century. It is now over 27 m (89 ft) tall, and almost as wide. The largest specimens, in Devon and in Surrey, have reached 30 m (98 ft) tall at the most.

The Lucombe oak was a chance hybrid between the cork oak (*Quercus suber*) and the Turkey oak (*Q. cerris*). It was raised at the nursery of Lucombe and Pince in Exeter, who were the first to introduce the Turkey oak to Britain in 1735. By 1762 their trees were fruiting, and they sowed the acorns. One of the seedlings proved to be different; it was extra vigorous and evergreen, at least until late spring. William Lucombe recognised that this was a hybrid and grafted as many trees as he could from the original onto Turkey oak stock and these grew to be the now impressive trees found around south Devon.

The story goes that when Lucombe was 90 years old, and his trees around 20 years old and 90 cm (3 ft) round, he had one cut down to make boards for his coffin, which he kept under his bed. These first boards became useless (perhaps they caught woodworm), so he had another tree felled and was buried in new planks in 1794, aged 102. The original trees grew very quickly, reaching around 18 m (60 ft) high in 27 years; one was illustrated in J. C. Loudon's *Arboretum et fruticetum Britannicum* (*The Trees and Shrubs of Britain*) published in 1838. Other seedlings were grown from the original cross, so there is considerable variation in the leaves of the different clones. A clone named 'Fulhamensis' is more common in the London area.

→ Tulip tree
Liriodendron tulipifera
Watercolour on paper
65 × 50 cm (26 × 20 in)

Tulip tree

The tulip tree, as it is called in Europe, or tulip poplar as it is known in North America, is one of the largest of all flowering trees. It is also ancient, belonging to the primitive family Magnoliaceae. There are two species – the more common *Liriodendron tulipifera*, which originates in eastern North America, and the rarer *Liriodendron chinense* that comes from China. Before the last ice ages, tulip trees were more widespread in the northern hemisphere and occurred in Europe and western North America as well. Their very characteristic leaves, which look as if they have been snipped off in a shallow V at the end, are found in fossil deposits in southern Germany, which were laid down around 34 million years ago in the Oligocene (Tertiary period), before the great Eurasian forests were killed off by the drier and colder conditions of the ice ages.

Early European settlers in America admired the tulip trees they discovered; the oldest recorded tree in England is said to have been planted in 1581 near Wheathampstead, Hertfordshire, and there are other recorded introductions through the 17th century. The biggest specimen at Kew is near the Azalea Garden, which was planted around 1770; it is now 35 m (115 ft) tall. Trees of similar size, around 30 m (98 ft) tall, are known from several places around Britain, but they are likely to be younger, and planted in the early 19th century. In North America the tulip poplar is found growing wild from Ontario to Louisiana, and can reach 50 m (164 ft) in sheltered places. Its straight bare trunks were used for making dugout canoes; its timber is known as yellow poplar, because of its pale colour. On both continents the young trees are fast growing, and can reach 17 m (56 ft) in 25 years. The young leaves are bright green, and the whole tree turns a buttery, golden yellow in the autumn.

The Chinese tulip tree is uncommon in cultivation and is becoming rare in the wild as it is felled for its timber. It occurs across a wide area, being recorded from Shaanxi and Hubei, south to Yunnan and North Vietnam, and can be found growing in rather damp valleys in the mountains. It is conspicuous for its reddish young leaves; its mature leaves are almost straight across at the apex, and its flowers, which are smaller than those of the American species, lack the orange markings. The trees reach 40 m (131 ft) in the wild, but the tallest specimens in cultivation are around 25 m (82 ft). This species was introduced into cultivation in Europe by the plant hunter E. H. Wilson in 1901, who discovered the species while he was looking for *Davidia* (see page 94). However, it has remained rare in cultivation. Trees grown from seeds collected by a Kew expedition to Daba Shan, in eastern Sichuan, have now been planted to form an avenue near the Azalea Garden at Kew.

Once young trees are established they are not difficult to grow, flowering around 20 years after planting. The seeds are dry and brown, with a narrow wing, unlike the fleshy seeds of *Magnolia*, which usually have a bright red skin and are embedded in a solid fruit.

↑ → Tulip tree, flowers and petals
Liriodendron tulipifera
Watercolour on paper
32 × 38 cm (13 × 15 in)

↑ → Tulip tree, leaf, flowers and petals
Liriodendron chinense
Watercolour on paper
25 × 38 cm (10 × 15 in)

→ Turner's oak
Quercus × *turneri*
Watercolour on paper
35 × 35 cm (14 × 14 in)

Turner's oak

Turner's oak (*Quercus* × *turneri*) is named after Spencer Turner (c.1728–76), who raised this hybrid tree at Holloway Down Nursery, in Essex. It is a cross between a holm oak (*Q. ilex*) and an English oak (*Q. robur*). Like most oak hybrids it was probably a mistake, appearing among seedlings of holm oaks that Turner was growing.

It is remarkable that this cross is so rare, as the two parent oaks often grow in close proximity and must be flooded with each other's pollen. However, there is considerable doubt as to how many different seedlings of this original cross were cultivated. Some distinguish Turner's original cross with the cultivar name 'William Turner', and Kew's tree as a second seedling called 'Pseudoturneri', because of small differences in leaf shape, the 'original' having shorter, broader leaves with mucronate (sharply-pointed) teeth. Others say that the two are from the same cross, and that the leaves can simply vary in shape.

Kew's tree was planted in 1798 near the present Princess of Wales Conservatory, and is presumably a graft of an original seedling. It is now a huge, spreading mass of dark green, wavy-edged evergreen leaves, which begin to turn brown and fall as the new leaves appear in spring. It often produces a good crop of acorns. It is remarkable that this tree survives at all, as it was blown up out of the ground by the Great Storm of 1987, and then dropped back into its hole, unscathed, though at a slight angle. Not only did it survive, but it grew better, as the compacted soil around the roots had loosened, allowing its roots to grow more freely. This led to a revolutionary treatment for many of Kew's old and poorly-growing trees, where the root area is injected with air and beneficial fungi, and a deep loose mulch is spread over the area of the root system to improve root health.

→ Corsican pine
Pinus nigra subsp. *laricio*
Watercolour on paper
52 × 40 cm (20 × 16 in)

Corsican pine

This grand specimen of a Corsican pine (*Pinus nigra* subsp. *laricio*) stands near Kew's Elizabeth Gate, next to Kew Green. It was planted in 1814 by R. A. Salisbury, a botanist, who collected it as a seedling in the south of France. Did the seed originate in Corsica or did Salisbury visit Corsica as well as France? Unfortunately we do not know. His pine has a great straight trunk and big slabs of pale brown bark, which are typical of this subspecies of the black pine.

The black pine (*Pinus nigra*) is generally divided into four subspecies, similar in general structure, but differing in minor characters and in distribution. The Austrian pine (subspecies *nigra*), is found mainly in central Europe; the Pyrenean pine (subspecies *salzmannii*) comes from the Pyrenees, Spain and North Africa; the Crimean pine (subspecies *pallasiana*) is widespread from Cyprus and Turkey to the Crimea, while the fourth – subspecies *laricio* from Corsica, southern Italy and Sicily is the one illustrated here. Many old specimens survive in deep valleys in the granite mountains of central Corsica, where they form wonderful forests of lofty trees, with straight, tall grey trunks.

The first Corsican pines were brought to Britain and Ireland around 1788, and there are other surviving specimens known to have been planted in the 1820s. They have grown best in the wetter, western parts of the UK, particularly in Scotland, where the tallest, at 45 m (148 ft), grows at Mount Stuart on the Isle of Bute. Other equally tall specimens are found in North Wales at Bodnant Gardens, and in Ireland at Powerscourt, south of Dublin.

Even though it may be the oldest in the country, the tree at Kew is only 30 m (98 ft) tall. The top was hit by a light aeroplane in 1928, and it has been struck by lightning at least twice, most recently in 1992. Pine trees, with their copious amounts of resin, are very sensitive to lightning. The resin boils in seconds, leading the timber to explode and set fire to the surrounding forest with its ground-cover of dry needles. Big areas of pine forest can be destroyed in this way. The resin, however, is also a valuable commodity; it is harvested by cutting in to the trunk and collecting in small pots or, nowadays, polythene bags. In ancient times Theophrastus reported that the mountain black pines were particularly valued for pitch, which was used widely as a glue and for waterproofing in shipbuilding, medicinally as a disinfectant, and among other things, as a cure for gout. Today it is used as 'rosin', for rubbing on the horsehair of bows to improve the vibration of the strings in cellos and violins, and as a source of turpentine, commonly used as paint thinner.

→ Stone pine
Pinus pinea
Watercolour on paper
67 × 97 cm (26 × 38 in)

Stone pine

According to Kew's Archives, this ancient specimen of the stone pine (*Pinus pinea*) was planted in the Gardens in 1846, about five years after Sir William Hooker came to Kew as its first director. It was at this time that he began to develop the royal gardens as a scientific collection. Another common name for *Pinus pinea* is the umbrella pine, as the trees usually form a rounded head, and from above (or on satellite imagery) a stone pine forest can look like a field of grassy mounds or a sea of green umbrellas.

The stone pine is found all around the Mediterranean, and there are still considerable forests in Spain and Portugal, in Italy and around Aydin in western Turkey; smaller stands are found elsewhere as far east as Georgia. Although it was known in England in the mid-16th century, when it was recorded in Turner's herbal, the big specimens in English gardens were all planted in the early 19th century. The tallest English specimens reach around 21 m (68 ft), for example at Dartington Hall in Devon, but stone pines can reach 35 m (115 ft) in the wild and in sheltered areas. This old Kew specimen was around 17 m (56 ft) tall, with low spreading stems, supposedly because it was grown as a pot specimen before being planted out. In its natural state it had a single, upright trunk and regular whorls of branches. The short greyish-green, rather stiff needles were in pairs, 8–12 cm (3–5 in) long; the large cones were rounded, with blunt scales, and stayed attached to the thick branches even after they had opened. This tree has now been removed. On a clear morning on 12th April 2023, the major trunk collapsed from the base, across the main path, and the remaining trunks were deemed too dangerous to save.

In addition to its beauty as a shade tree, the stone pine is often cultivated for its seeds, which are sold as pine nuts – an ingredient of pesto and common in baklava. They are known to have been used extensively by the Romans, and carbonized nuts have been found in the ruins of Pompeii. Other species of pine also have edible nuts. *Pinus koraiensis* nuts are commonly exported from eastern China, and seeds of *P. armandi* and *P. bungeana* are also eaten. Several species of pinyon pine nuts have been harvested in western North America where they were a staple food of the Native Americans.

↑→ Stone pine, cone
Pinus pinea
Watercolour on paper
70 × 50 cm (28 × 20 in)

→ Chestnut-leaved oak
Quercus castaneifolia
Watercolour on paper
85 × 72 cm (33 × 28 in)

Chestnut-leaved oak

This specimen of the chestnut-leaved oak (*Quercus castaneifolia*) is certainly one of the most magnificent trees in the whole of Kew, as well as being one of the largest of its kind in Britain. Its leaves resemble those of the sweet chestnut but the acorns' bristly cups are typically oak-like, and very similar to those of the common Turkey oak (*Quercus cerris*).

This rare oak was collected and described by Carl Anton Meyer from the Talysh Mountains in Azerbaijan. Mayer spent the winter and spring of 1829–30 in Baku, and in June travelled through the thick forests on the northern slopes of the Talysh, and it was here he saw *Quercus castaneifolia*, as well as the large-leaved alder, *Alnus subcordata*, which also thrives at Kew. *Quercus castaneifolia* has since been found from the eastern Caspian near Bojnurd, west to the border of Iran and into Azerbaijan, in the Talysh, growing in the humid forests that curve round the southern coast of the Caspian Sea. This is an area where several trees and woodland plants which were widespread in the late Tertiary forests managed to survive the ice ages, which began around 1.8 million years ago. *Parrotia persica*, *Acer velutinum*, *Gleditsia caspica*, *Zelkova carpinifolia* and *Pterocarya fraxinifolia* all have other living relatives in China.

Quercus castaneifolia generally grows in humid areas at low altitude up to 2,000 m (6,562 ft) on the north side of the mountains, sometimes, as in the Gulestan forest west of Bojnurd, forming nearly pure stands of trees around 300 years old.

Kew's Champion Tree of *Quercus castaneifolia*, which is illustrated here, was probably planted under the direction of Sir William Hooker in 1846, as part of his extension of the arboretum. It appears to have come from seed collected in 1843; the collector is not known. When measured in 1909 it was 18 m (59 ft) high and 70 cm (2 ft) in diameter, while in 2009 its height was 39 m (128 ft) with a diameter of 245 cm (8 ft). The largest specimens can reach 50 m (164 ft) in height. Another smaller but still massive specimen of *Quercus castaneifolia*, with branches that sweep the ground, can be seen near the Orangery. It is much younger, only planted in 1953, and is already over 20 m (66 ft) high. These great oaks are pruned regularly; around a quarter of the leaf canopy is removed every five years, and this has helped them to survive several major storms including the great storm of 16th October 1987 and the damaging gales of winter 2013–14.

Acorns collected from the chestnut-leaved oaks at Kew germinate well if planted immediately, but are likely to be hybrids with related species of oak growing at Kew, usually the Turkey oak, which has deeply-divided leaves. These hybrids are very fast growing and can put on a metre of growth per year in ideal conditions. An upright form of *Q. castaneifolia* 'Green Spire' is more compact, and is an admirable tree for planting to form an avenue or a single specimen where space is limited.

↗ Chestnut-leaved oak
Quercus castaneifolia
Watercolour on paper
56 × 45 cm (22 × 18 in)

→ Giant sequoia, cone
Sequoiadendron giganteum
Watercolour on paper
24 × 17 cm (9 × 7 in)

Giant sequoia and coast redwood

The giant sequoia, sierra redwood, or Wellingtonia as it is commonly called in England, is the most massive temperate tree on Earth. The huge old specimens that survive in California contain as much as 1,487 cubic metres (52,513 cubic feet) of wood. *Sequoiadendron giganteum* was observed in 1833 by J. K. Leonard, and mentioned in his diary, but the report did not reach the botanical establishment. Another early explorer, John M. Wooster, carved his initials on a particularly large tree, known as Hercules. This was blown down in 1861, but remains at its original location. The next sighting was in 1852, by Augustus Dowd, a hunter employed to feed the miners in the gold rush. He discovered it by chance, as he was following a wounded grizzly bear near Calaveras, and hailed himself as the discoverer. This grove, the North Grove, and the nearby South Grove are now part of a state park. The biggest survivor of all, however, is in Tulare County. This is the tree called General Sherman, now around 84 m (276 ft) high, 11 m (36 ft) in diameter at the base and 2,500 years old.

Sequoiadendron was introduced to cultivation in England by William Lobb, a very successful plant collector, who was then working for Veitch's Nursery at Coombe Wood in Surrey; he heard of Dowd's discovery and rushed to collect seed (and two seedlings), which he brought to England via Mexico, as he thought that was the quickest route. Unknown to him, a Scot, John D. Matthew from Geordiehill near Errol in Perthshire, also heard of Dowd's discovery of the tree and brought back seed to Scotland, aboard the first steam packet to cross the Atlantic, arriving in August 1853. Matthew shared seed among the neighbouring lairds, and the surviving trees are probably the oldest in Europe. A tree at Castle Leod, said to have been planted in 1853, is now 52 m (171 ft) tall, and there is a slightly taller one at Blair Castle nearby. Lobb's introductions have also grown well, but have not reached the height of those in the sheltered Tay Valley. Kew's specimens date from the 1860s, and the tallest is now 36.5 m (120 ft) high. →

→ Coast redwood, leafy shoot
Sequoia sempervirens
Watercolour on paper
24 × 17 cm (9 × 7 in)

→ All over England giant redwoods stand out above other trees, their dark cigar shapes dominating the skyline. Often the top is killed by lightning, and one or more side branches continue upwards. Only in the most sheltered places, in deep valleys, can the trees be expected to reach their full height. The name Wellingtonia was given to the tree by John Lindley, secretary of the Horticultural Society and a noted botanist. He named it in honour of Arthur Wellesley, the first Duke of Wellington, and hero of the battle of Waterloo. Fine specimens, planted in the late 1850s, are still growing on his estate at Stratfield Saye, which was given to him by the nation after the battle.

If the Wellingtonia or sierra redwood is the most massive temperate tree on Earth, the coast redwood, *Sequoia sempervirens*, is the tallest. At nearly 116 m (380 feet), the tallest specimen, called Hyperion, is in Redwood National Park in Humboldt Co., northern California. Taller trees are known, but are no longer standing. The coast redwood is easily recognised by its flat spray of leaves, and very soft, reddish, spongy bark. In the forests in California, the huge trees are remarkably close together, their trunks soaring up into the fogs that roll in from the Pacific, cooled by the cold, southward-flowing current. It is these humid summer fogs, combined with winter rainfall, that allow the redwoods to grow so tall.

Coast redwoods are easily grown in north-western Europe, and have already reached 50 m (164 ft) since they were first planted in the late 1840s. The tallest trees are in the western fringes of Europe, in Devon, in North Wales (at Bodnant), in Scotland and in Ireland, growing in deep, moist, sheltered valleys. Young trees grow fast, reaching around 30 m (98 ft) or even 40 m (131 ft) in fifty years. The first trees were brought to Britain not, as might have been expected, by David Douglas, whose 1833 collections were lost when his canoe was swamped, but through the St Petersburg Botanic Garden, originally collected near the Russian station at Fort Ross in Sonoma Co. The director of the garden, Dr Fischer, sent plants to Knight and Perry's London nursery in 1844. In 1850, larger quantities of seed were sent back by William Lobb who was collecting for Veitch's nursery in Exeter.

Trees of the original Russian introduction are still growing in several grand gardens, such as Dropmore in Buckinghamshire, Longleat in Wiltshire and Stratfield Saye, but have been overtaken in height by the later plantings. One of the most beautiful sites is a grove of twenty towering redwoods in the garden at Whitfield in Herefordshire.

→ Armand's pine
Pinus armandi
Watercolour on paper
53 × 73 cm (21 × 29 in)

Armand's pine (Chinese white pine)

Armand's pine (*Pinus armandi*), also known as Chinese white pine, is one of the elegant Himalayan pines, recognised by their long, slender needles in bunches of five and elongated cones, which hang from the branches and often drip with resin. The very smooth green bark of the young trunk and branches are characteristic of this species. It is found through most of western and southern China, growing on mountain slopes in wet areas. The first European to discover it was the missionary plant hunter Père Armand David (see page 94) in late 1873 in the Tsinling Shan of Shensi. It was scientifically named in 1884, but not introduced to cultivation until 1895. The largest specimens are in Ireland, and one at Fota House, County Cork, had reached 29 m (95 ft) tall in 1999 – nearing the maximum of 35 m (115 ft) indicated in the *Flora of China*.

There are several new plantings of this tree at Kew and at Wakehurst, grown from seed collected in China since 1980. They are now beautiful young trees, and it is from one of these that this painting was made. Like the Bhutan pine (*Pinus wallichiana*), which is closely related, *P. armandi* is fast growing but not very long-lived, requiring both moisture and good drainage. Old specimens are rare, though the one at Fota is recorded as being planted in 1917.

Pinus armandi is one of several species cultivated for pine nuts, particularly in China, and has been implicated in causing pine nut syndrome, in which everything tastes bitter; although not all varieties of *P. armandi* seem to have this effect, and it only affects some people. More information is needed on the causes of the syndrome.

→ Handkerchief tree
Davidia involucrata
Watercolour on paper
50 × 70 cm (20 × 28 in)

Handkerchief tree (dove tree)

Davidia involucrata, known as the handkerchief tree but also called the dove tree and ghost tree, is one of the most elegant of trees, especially when adorned with its delicate, white flowers, which hang down from horizontal branches. The white parts of the flowers are actually a pair of thin, papery, leaf-like bracts, which surround a tight cluster of the actual flowers: numerous male flowers consisting mainly of stamens with dark anthers, and one or two slightly longer female flowers. In late summer the branches have green fruits hanging on long stems, with a hard, ribbed woody nut inside. This holds up to four seeds.

Davidia is named after the French naturalist and missionary Père Armand David, who encountered it in western China in 1869 but did not bring seed to Europe. The trees grow in deep valleys in the mountains and are very rare, usually occurring as single trees in dense, wet forests. The story of its introduction to cultivation is one of the most convoluted in botanical history. After David, the next botanist to see *Davidia* was the Irishman Augustine Henry, who reported seeing one tree in flower in 1888, near Wushan in eastern Sichuan. In 1899 the great nurseryman Sir Harry Veitch sent to China a young man from Gloucestershire, Ernest H. Wilson, with instructions to find Henry's tree and bring back seed. By then Henry, who worked for the Chinese customs service, had been posted to Simao in southern Yunnan, near the Vietnamese border, so Wilson crossed China to meet Henry, and then went back to central China again via Hanoi, Hong Kong and Shanghai, armed with a sketch map to help locate Henry's tree. Wilson traced the locality, only to find that the tree had been cut down the year before to make the owner's house. Undaunted, as he proved to be on several other occasions, Wilson found other trees near Ichang, and sent back such quantities of fruits that they resulted in no less than 13,000 plants. On his return he discovered that another French Missionary, Père Farges, had already sent seed to M. Maurice de Vilmorin, who had passed on cuttings to Kew and the Arnold Arboretum in Boston, Massachusetts. These introductions were of the variety *vilmoriniana*, which has hairless leaves.

The variety that David found originally has larger leaves that are silky-hairy underneath, and was introduced by Wilson in 1903 and 1904. Both varieties are represented in the Kew collection, but not as old trees. In fact, few trees of the thousands raised from Wilson's seeds have survived; two of Wilson's originals at Hergest Croft, Herefordshire have reached 20 m (66 ft) in height, as have a few old trees in other gardens, such as Dartington Hall in Devon and Bodnant in North Wales. A very large wide-spreading tree at Rowallane, County Down, is a cutting from Père Farges' introduction. The Arnold Arboretum, where Wilson worked in later life, also has one tree from his collection.

Davidia is not difficult to grow, but is susceptible to drought and needs a wet site to grow well. Excellent specimens are also found in the University Botanic Garden in Cambridge, where the soil, although chalky, is watered by spring-fed streams.

→ Indian horse chestnut
Aesculus indica 'Sydney Pearce'
Watercolour on paper
40 × 50 cm (16 × 20 in)

Indian horse chestnut

When in full flower, the Indian horse chestnut (*Aesculus indica*) is one of the most spectacular trees at Kew. The three individuals planted in a row, just inside the entrance from Kew Green, become covered with upright pyramidal clusters of white and pink flowers, and are a sight to behold. It is surprising that this beautiful tree is not planted more often, as it is very hardy and free-flowering, as well as being healthy; the leaf mining moth, *Cameraria ohridella* does not seem to attack it. The Indian horse chestnut was introduced into cultivation in 1851 by Col. Henry William St Pierre Bunbury, who collected the seeds while serving in the Indian army on the frontier of India and Afghanistan; he had previously explored parts of Western Australia, where he gave his name to the city of Bunbury.

Bunbury's chestnut trees grew very fast and first flowered 1855 or 1856 in the garden of his father, Sir Henry Edward Bunbury at Mildenhall in Suffolk. His brother, C. J. Fox Bunbury Esq., an amateur botanist who collected plants in South Africa, South America and Tenerife, sent specimens to Kew's director, Sir William Hooker, and they were illustrated in *Curtis's Botanical Magazine* in 1859.

The Indian horse chestnut grows wild between 1,800 m and 3,000 m (5,905 ft and 9,842 ft) elevation in the mountains from Afghanistan to central Nepal, growing in moist, sheltered valleys. The locals use the bark to treat rheumatism, and eat the seeds in times of famine, after soaking them to remove the bitter saponins. They are given to horses to treat colic, and the leaves are also cut as fodder.

The finest of the Kew trees has been given the cultivar name 'Sydney Pearce'. Sydney was a student at Kew in 1928 and then returned to Kew as a member of staff, becoming assistant curator of the Arboretum in 1937, and then deputy curator from 1967 onwards. The tree which is named after him was raised from seed by C. F. Coates in 1928 from an old tree at Kew, which had been planted in 1887, most likely itself a seedling from Bunbury's tree. It is a particularly free-flowering tree with large flowers of a good colour.

Masumi Yamanaka is particularly impressed by this tree, and has drawn it at all stages in its life history, from conker and seedling, to the breaking buds, flowers, fruit and autumn colour. It is notable that the flowers open white, and become pinkish after the pollen has been shed. Seeds from Kew's tree grow well, but are not often as free-flowering as their parent.

↓ Indian horse chestnut, fruits
Aesculus indica 'Sydney Pearce'
Watercolour on paper
18 × 23 cm (7 × 9 in)

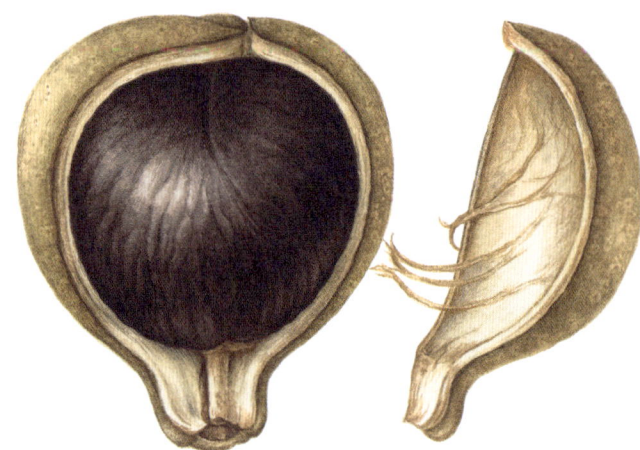

↓ Indian horse chestnut, germination
Aesculus indica 'Sydney Pearce'
Watercolour on paper
38 × 22 cm (15 × 9 in)

↓ Indian horse chestnut, early spring
Aesculus indica 'Sydney Pearce'
Watercolour on paper
27 × 37 cm (11 × 15 in)

↓ Indian horse chestnut, early summer
Aesculus indica 'Sydney Pearce'
Watercolour on paper
50 × 42 cm (20 × 17 in)

↙ Indian horse chestnut, summer
Aesculus indica 'Sydney Pearce'
Watercolour on paper
43 × 51.5 cm (17 × 20 in)

↓ Indian horse chestnut, early autumn
Aesculus indica 'Sydney Pearce'
Watercolour on paper
32 × 26 cm (13 × 10 in)

↓ Indian horse chestnut, autumn
Aesculus indica 'Sydney Pearce'
Watercolour on paper
40 × 50 cm (16 × 20 in)

→ Bhutan pine
Pinus wallichiana
Watercolour on paper
27 × 19 cm (11 × 7 in)

Bhutan pine

Bhutan pine (*Pinus wallichiana*) is the most widespread five-needled pine, both in its native Himalayas, and in cultivation in gardens. It is fast growing when young, with fine bluish-green needles hanging from the branches. The cones are long, up to 30 cm (1 ft), hanging in clusters of two or three, and dripping with sticky white resin.

In the wild *Pinus wallichiana* is found in dry forests, ranging from eastern Afghanistan to southeast Tibet (Xizang), western Yunnan and northern Burma, and growing at altitudes between 1,800 m and 4,000 m (5,906 ft and 13,123 ft). Here it makes a tall tree, to 50 m (164 ft), forming a dark-grey bark from its second year. The needles are particularly blue when young, and can be 20 cm (8 in) long. The rare *Pinus bhutanica,* from the forests of Bhutan, has five longer needles, which are even more drooping and can be 28 cm (11 in) long.

Pinus wallichiana grows well in Britain, particularly where the rainfall is high. Exceptionally well-grown trees can reach 20 m (66 ft) in 60 years. For the first 20 years or so the trees have a leader and make an upright trunk, but after this age the top branches spread out and the tree becomes more rounded and untidy. The old specimen at Kew, near the Palm House, was planted in 1846 and is a somewhat craggy tree, with hanging branches that fruit very freely. It reached 26 m (85 ft) in height, but is now in decline.

The oldest surviving tree in the UK was planted in 1831, soon after the introduction of this species in 1823 by A. B. Lambert. Lambert reports that he grew many young trees from seed, but it is not clear who collected the seed; it may have been Nathaniel Wallich, who is mentioned as supplying Lambert with numerous specimens of this pine from Bhutan. Wallich was a great plant collector and botanist, and superintendent of the botanic garden in Calcutta; in the early 19th century he sent a great many plants back home and, in 1834, was involved in the exploration of tea plants growing wild in Assam.

Other old specimens of *Pinus wallichiana* may be from this first introduction. In Ireland, for example, there are two old specimens at Powerscourt in County Wicklow, and one at Fota House in County Cork that was planted in 1847 is now around 30 m (98 ft) tall. The tallest known specimen grows near Welshpool, in mid Wales, and was measured at 35 m (115 ft) in 2009; others nearly as high are recorded in western Scotland.

Pinus wallichiana has gone through several name changes, both in Latin and in English. Old books often call it *Pinus excelsa*, a name published in 1828 by David Don in the second edition of Lambert's beautiful monograph of *Pinus*, but abandoned because Lamarck, in his *Flore Français* in 1759, had already given this name to a different species. A later name, *Pinus griffithii,* was published in 1854 and commemorated William Griffith who collected in Afghanistan and the Himalaya for the East India Company. The current name *Pinus wallichiana* was used by A. B. Jackson

in the *Bulletin of Miscellaneous Information*, commonly called *Kew Bulletin*, and acknowledges that Nathaniel Wallich was the first to name this pine. Jackson was a noted authority on conifers and with William Dallimore, foreman in the Arboretum at Kew, published the classic *Handbook of Coniferae* in 1923. Its English names have also been changed: Bhutan pine is commonly used but can be confused with *Pinus bhutanica*, furthermore, *P. wallichiana* grows well beyond the borders of Bhutan. Blue pine is a common name in India. Himalayan blue pine is the name given by Oleg Polunin and Adam Stainton in their *Flowers of the Himalaya* in 1984.

→ Nikko maple
Acer maximowiczianum
Watercolour on paper
46 × 38 cm (18 × 15 in)

Nikko maple

The Nikko maple, now called *Acer maximowiczianum*, but formerly called *A. nikoense*, forms a sturdy tree with rather thick, hairy twigs and leaves with three leaflets which are shallowly lobed, rather like some oaks. In Japan, this maple is found in the mountains of Honshu, Kyushu and Shikoku, but is not common anywhere. In China it is recorded from western Hubei, eastern Sichuan and neighbouring states; Chinese trees are bigger and hairier in leaf and fruit, and were called var. *megalocarpum* by Alfred Rehder when describing E. H. Wilson's specimens from Hubei. Its main attraction for gardeners is the bright red and yellow colour of the autumnal leaves. In Japan the bark is used to make eye drops.

The flowers, which are shown opposite, are around 1.2 cm (0.5 in) across, and are functionally unisexual. The two male flowers (shown left), have numerous well-developed stamens, plenty of pollen and little trace of an ovary; the two female flowers, shown right, have an ovary, two distinct styles and poorly-developed stamens which usually lack pollen. The paired and winged fruits are typical of all species of maple.

In the wild the trees may reach 20 m (66 ft), but they are slow growing; the largest in cultivation are at Westonbirt Arboretum and only around 18 m (59 ft) tall. There are other big specimens at Hergest Croft Gardens in Herefordshire and at East Bergholt Place in Suffolk. This species is rarely seen outside arboreta.

The species name honours Carl Johann Maximowicz, a Russian botanist who came across the tree growing in Japan in the 1860s, while its common name refers to the Japanese town of Nikko. *Acer maximowiczianum* was introduced to cultivation in England in 1881 by Charles Maries when he was collecting in the forests of Hokkaido, Japan, for Messrs Veitch, who then had a nursery at Coombe Wood, near Kingston-on-Thames.

The Chinese *Acer griseum* is very similar to *A. maximowiczianum*, particularly in leaf, but is easily recognised by its smooth, peeling, reddish bark and thinner twigs, which are smooth, not hairy.

→ Indian bean tree
Catalpa bignonioides
Watercolour on paper
36 × 28 cm (14 × 11 in)

Indian bean tree (southern catalpa)

Catalpa bignonioides is often called the southern catalpa to distinguish it from the very similar western catalpa (*C. speciosa*). Both are from the south-eastern United States, and are rather rare in the wild. *C. bignonioides* is native to Alabama and Mississippi, as well as small areas of Florida, Georgia and Louisiana, while *C. speciosa* grows along the Mississippi River in Indiana, Tennessee, Ohio and Missouri. Both species flower in July and August and form wide-spreading, large-leaved trees, which grow quickly and have thick, very pithy twigs and soft wood. *C. speciosa* is the taller tree with fewer, larger flowers in each panicle, and the leaves smell acrid when crushed.

After flowering, long hanging seed pods are formed, leading to its alternative name, the Indian bean tree, as the word catalpa is the Cherokee name for a bean. The flat, winged seeds indicate that *Catalpa* belongs to the family Bignoniaceae, which contains many climbers, including the red-flower climber *Campsis*, also from the south-eastern USA. *Catalpa* is found only in eastern North America and China. *C. fargesii* is found mainly in western China and *C. bungei* in the northeast around Beijing, both bear pink or purplish flowers in the spring. *C. ovata* has yellowish flowers, and a long history of cultivation in both China and Japan. *C. tibetica*, with white flowers, is a small tree from high altitudes in Yunnan and Tibet. Though the first three are planted as street trees in China, they are much rarer in cultivation in Europe and North America than the two American species.

As they originate in warm parts of North America catalpas need a warm summer to flower well, and they are especially floriferous in the extra heat of cities such as London and Paris. In the cooler parts of Europe they do not flower so well. The mature specimens at Kew are planted near the Temperate House, and generally flower freely, setting ample fruit.

The first recorded introduction of *Catalpa* to Europe was in 1726, when Mark Catesby sent seed to England, as well as growing them himself in Carolina. Indian bean trees are not long-lived. They grow fast to 10 m (33 ft) or very rarely 20 m (66 ft), and then begin to die slowly. Even on the well-drained, gravelly soil at Kew, the finest specimens have lived for only 60 years, and others are known to have had even shorter lives.

Goat horn tree
Carrierea calycina
Watercolour on paper
40 × 48 cm (16 × 19 in)

Goat horn tree

Carrierea calycina, sometimes called the goat horn tree because of its strange, horn-like fruits, is very rare in cultivation, and has seldom been collected in the wild. It is native to China, and was first brought into cultivation in England by E. H. Wilson, who collected it at around 1,000 m (3,281 ft), both in Hubei and Sichuan, between 1903 and 1908. It has now been recorded from Yunnan, Guizhou, Guangxi and Hunan as well. Wilson's trees did not thrive in cultivation, though it flowered in a few gardens around 1930, and this species soon became very rare. One of the few that survived grows at Birr Castle in County Offaly, Ireland and has reached 17 m (56 ft) in height. Wilson mentions the edges of streams as a frequent habitat, so it is likely that the tree will not tolerate summer drought.

In October 1994 seeds of *Carrierea* were collected by Peter Wharton, the late curator of the David C. Lam Asian garden of the UBC Botanic Garden in Vancouver, and most of the trees in cultivation are derived from this collection. Wharton collected it in the Dashahe Nature Reserve in the Dalou Shan in northern Guizhou, where it formed a flat-topped tree, growing by rivers or at the foot of north-facing cliffs; several seedlings are growing at Kew and Wakehurst, the earliest of which were planted in 1995. The tree from which this specimen was painted was growing in the garden of the late Harry Hay near Reigate, Surrey, who received some of Peter Wharton's seed through Roy Lancaster. The resulting seedlings first flowered in Roy Lancaster's garden in 2004, and in Harry Hay's garden in July 2006.

Two species of *Carrierea* are recognised in the *Flora of China*: *C. calycina*, which was discovered by the Abbé Farges near Baoxing, and the rarer *C. dunniana,* which is a smaller shrub with smaller leaves and flowers, and a more southerly distribution. Both have flowers with whitish sepals and no petals; male and female flowers are generally produced on different trees. *Carrierea* was traditionally placed in the family Flacourtiaceae, but DNA studies of this and related families have resulted in most of the genera of old Flacourtiaceae being placed in Salicaceae, despite their superficial lack of similarity to willows (*Salix*) and poplars (*Populus*).

→ Bogong gum
Eucalyptus chapmaniana
Watercolour on paper
43 × 62 cm (17 × 24 in)

Bogong gum

The Bogong gum (*Eucalyptus chapmaniana*) is native to New South Wales and Victoria, where it grows in the mountains. Its local name derives from its locality, as it grows near Mount Bogong, the highest peak in the Australian Alps of Victoria.

In the wild *Eucalyptus chapmaniana* can reach 30 m (98 ft) in height, with a dense, rounded top, and this is also one of the features of the tree in cultivation. The bark is brown and cracked on the main trunk, but hangs in ribbons off the branches. The flowers are white, measuring about 1.8 cm (0.7 in) across, with the fruits a little over 1 cm (0.4 in) across. These flowers are one of the distinctions from the similar, much commoner and equally hardy *Eucalyptus dalrympleana*, the mountain gum, which has flowers only 1.5 cm (0.6 in) across.

Eucalyptus chapmaniana is named after Brigadier Wilfrid Chapman (1891–1955), an eminent civil engineer, who was born in Wandsworth, England, and emigrated to Australia when his father was appointed paleontologist to the National Museum in Melbourne. Chapman served in both World Wars, and became a pioneer in the use of electric arc welding, as well as being a keen amateur botanist.

This is one of the superb *Eucalyptus* trees growing near the Children's garden at Kew. It was planted in 1988, and is now around 16 m (52 ft) tall; a tree from the same source at Wakehurst is bigger and was measured at 24 m (79 ft) tall in 2010. You can also find other splendid specimens of *Eucalyptus* in Kew, near Brentford Gate.

→ Sapphire dragon tree, fruits
Paulownia kawakamii
Watercolour on paper
40 × 30 cm (16 × 12 in)

Sapphire dragon tree

When covered with large dark-spotted, pale lilac flowers, the sapphire dragon tree (*Paulownia kawakamii*) is one of the most beautiful of all trees. It was introduced to Kew in 1992, after seeds were collected by Tony Kirkham and Mark Flanagan on a Kew collecting expedition in Taiwan. Even in its native country this is a very rare tree, perhaps with only a hundred individuals left in the wild. At Kew it has grown very quickly, and proved surprisingly hardy, since it was planted out in April 1995. It does not flower every year, which is not surprising as in good years every branch ends in a pyramid of over a hundred flowers, each three to five centimetres across, and the tree takes a year or two to build up to flowering strength again.

The genus *Paulownia* was named by Franz von Siebold, to honour Princess Anna Paulowna, daughter of Czar Paul I of Russia; she was born in 1795 in St Petersburg, and married the Prince of Orange, later King William II of the Netherlands. Siebold found the common *Paulownia tomentosa* growing in Japan, and both there and in China its very light wood is used for making furniture.

There is a tradition in Japan that a *Paulownia* tree is planted when a daughter is born, and then used to make a chest for her trousseau when she gets married. Young trees are so fast-growing that in less than 20 years they can have a substantial trunk.

Many different species of *Paulownia* have been named, all growing in China, but the current opinion is that there are seven species, two of which are found in Taiwan. The genus has no near relatives, and is placed in its own family; it is probably closest to the mostly parasitic Orobanchaceae, or to the Phrymaceae, which includes *Mimulus*. *Paulownia kawakamii* can be recognised by the combination of its very large, branching clusters of flowers, deeply-lobed calyx and very short flower stalks, but it is the wide-mouthed, densely spotted foxglove-like flowers that are most striking. It is probably truly wild only in Taiwan, but it is found as a cultivated tree in many parts of mainland China. Paulownias are prolific seeders and easily escape into the wild, as has *P. tomentosa* in eastern North America, so their true native distribution, particularly within China, is now uncertain.

↓ Sapphire dragon tree, flower
Paulownia kawakamii
Watercolour on paper
40 × 30 cm (16 × 12 in)

Recommended reading

Bean, W. J. (1970–80) *Trees and Shrubs hardy in the British Isles.* 8th ed., revised by D. L. Clarke. 4 vols. John Murray, London.

Desmond, R. (2007) *The History of the Royal Botanic Gardens, Kew.* Royal Botanic Gardens, Kew.

Elwes, H. J. & Henry, A. (1906–13) *The Trees of Great Britain and Ireland.* 8 vols. Privately printed.

Flanagan, M. & Kirkham, T. (2009) *Wilson's China.* Royal Botanic Gardens, Kew.

Fry, C. (2012) *The Plant Hunters: The Adventures of the World's Greatest Botanical Explorers.* Andre Deutsch, London.

Grimshaw, J. & Bayton, R. (2009) *New Trees.* Royal Botanic Gardens, Kew.

Harrison, C. (2020) *The Botanical Adventures of Joseph Banks.* Royal Botanic Gardens, Kew.

Harrison, C. (2019) *Kew's Big Trees.* 2nd edition. Royal Botanic Gardens, Kew.

Harrison, C. & Kirkham, T. (2024) *Remarkable Trees.* Thames & Hudson.

Hemrey, G. & Simblet, S. (2014) *The New Sylva: A Discourse of Forest and Orchard Trees for the Twenty-First Century.* Bloomsbury Publishing, London.

Hillier, J. G. & Lancaster, R. (2014) *The Hillier Manual of Trees and Shrubs.* 8th edition. Royal Horticultural Society, London.

Hobhouse, P. (1999) *Plants in Garden History.* Pavilion Books, London.

Johnson, O. (2011) *Champion Trees of Britain and Ireland: The Tree Register Handbook.* Royal Botanic Gardens, Kew.

Kilpatrick, J. (2014) *Fathers of Botany.* Royal Botanic Gardens, Kew.

Kirkham, T. (2022) *Arboretum.* Big Picture Press.

Meiggs, R. (1982) *Trees and Timber in the Ancient Mediterranean World.* Oxford University Press, Oxford.

Mitchell, A. (1996) *Alan Mitchell's Trees of Britain.* Collins, London.

Ogilvy, R. & Ogilvy, S. (2013) *Overleaf.* Royal Botanic Gardens, Kew.

Pakenham, T. (1997) *Meetings with Remarkable Trees.* Weidenfeld and Nicolson, London.

Stokes, J. (2002) *Great British Trees.* The Tree Council, London.

Willis, K. & Fry, C. (2014) *Plants: From Roots to Riches.* John Murray, London.

Wilson, E. H. (1911–16) *Plantae Wilsonianae: An enumeration of the woody plants collected in western China for the Arnold Arboretum of Harvard University during the years 1907, 1908 and 1910*, edited by C. S. Sargent. 3 vols,. University Press, Cambridge. Reprinted by Dioscorides Press (1988).

Wulf, A. (2009) *The Brother Gardeners.* Windmill Books, London.

Websites

Flora of China (online). http://flora.huh.harvard.edu/china/

www.globaltrees.org

www.treesandshrubsonline.org

www.treeregister.org

Artist and authors' biographies

Masumi Yamanaka was born in Nara, Japan in 1957, and now splits her time between the UK and Japan. She came to the UK in 1987 as a ceramics designer for Marks & Spencer and has worked in this capacity for manufacturers and retailers such as Royal Doulton, Portmeirion, The National Trust, Vista Alegre (Portugal) and Crate & Barrel (USA).

Masumi studied botanical art under Pandora Sellars and has been a freelance artist based at the Royal Botanic Gardens, Kew since 2007. She is also Japanese exhibition coordinator for Kew.

Her work shown at the Royal Horticultural Society has included a series of camellias, for which she was awarded a Silver-Gilt Medal, and her Gold Medal award-winning illustrations of *Aesculus indica* 'Sydney Pearce', illustrated from the specimen at Kew. She was presented with The Margaret Granger Memorial Silver Bowl by the Society of Botanical Artists for her watercolour illustration of 'Hippeastrum'. This illustration appears in the SBA's associated publication *The Botanical Palette* and the original work is part of the collection at Kew.

Masumi's works have been exhibited regularly at Kew's Shirley Sherwood Gallery of Botanical Art and are held at Kew and in private collections throughout the world.

Christina Harrison is the former editor of the award-winning *Kew* magazine (2008–20). She has a first degree in Ecology and an MA in Garden History with a dissertation on the history of Kew's trees. She has published several books on trees and plant history, and is currently undertaking a PhD to research the history of Kew's landscape and its tree collections with Royal Holloway, University of London. She is a fellow of the Royal Historical Society, Royal Geographical Society and Linnean Society.

Martyn Rix is a botanist and was editor of *Curtis's Botanical Magazine* from 2003–2024, with an unrivalled knowledge of botanical art. With a PhD in Botany from the University of Cambridge, he has worked at the University Botanic Garden, Zurich and at the RHS Garden, Wisley.

He is a life member of the International Dendrology Society, and has made many expeditions to different parts of the world to study trees and other plants. He is the author of a number of books, including the highly acclaimed *Rory McEwen, the Colours of Reality* and *The Golden Age of Botanical Art*.

Index

Italic numbers indicate illustrations

A New Herball (Turner) 33
Acer griseum 110
Acer maximowiczianum 110
Acer negundo 24
Acer velutinum 82
Aesculus indica 'Sydney Pearce' 10, *14*, *19*, *32*, 98, *99*, *100–1*, *102–3*, *104–5*, *106–7*
Aiton, William 27, 44
Alnus subcordata 82
American plane 24
Arboretum et fruticetum Britannicum (Loudon) 62
arboretum, Fulham Palace 24
Arboretum, Kew 7–8, 10, 28, 30
   carbon capture 31
   champion trees 8, 10, 22, 28–9, 30, 82
   environmental benefits 30–1
   first garden 27
   Great Storm, 1987 7, 70, 82
   heritage trees 10, 22
   Office of Woods and Forests 28
   'Old Lions' 7, 27, 48
   pinetum 28
   plant boxes *27*
   plant hunters 28
   storm of 1916 52
   tree collectors 22–4, 30
Argyll, Duke of 24, 52, 58
Armand's pine 90, *91*, *92–3*
Atlas cedar 28
Augusta, Princess 10, 22, 27, 44, 52, 58
Austrian pine 74
Banks, Sir Joseph 27–8
Bartram, John 25
beech 26
Bhutan pine 90, 108, *109*
black locust tree 24, 27, 52, *53*, *54–5*, *56–7*
black pine 74
black walnut 30
Bogong gum 116, *117*

Bridgeman, Charles 24–5, 26, 34
Brompton Park Nursery 24
Brown, Lancelot 'Capability' 25
Bunbury, Col. Henry William St Pierre 98
Bute, Lord 26, 27, 28
Cameraria ohridella 98
Campsis 112
carob 52
Caroline, Queen 25
Carrierea calycina 114, *115*
Carrierea dunniana 114
Castanea sativa 26, 34, *35*, *36–7*
Catalpa bignonioides 112, *113*
Catalpa bungei 112
Catalpa fargesii 112
Catalpa ovata 112
Catalpa speciosa 112
Catalpa tibetica 112
Catesby, Mark 112
Cathaya argyrophylla 48
cedar of Lebanon 38, *39*, *40–1*, *42–3*
Cedrus atlantica 28
Cedrus deodara 28
Cedrus libani 38, *39*, *40–1*, *42–3*
Cedrus libani subsp. *brevifolia* 38
Ceratonia siliqua 52
Chambers, Sir William 27, 52
Chapman, Brigadier Wilfrid 116
Chelsea Physic Garden 25
chestnut-leaved oak 8, *18*, 28, 82, *83*, *84–5*
Chinese white pine 90, *91*, *92–3*
coast redwood 28, *89*, 88
Coates, C.F. 98
Collinson, Peter 25
Compton, Henry, Lord Bishop of London 24
cork oak 62
Corsican pine 74, *75*
Crimean pine 74
Cunninghamia lanceolata 48
Cupressus sempervirens 23
Curtis's Botanical Magazine 98
David, Père Armand 90, 94

Davidia involucrata 29, 94, *95*, *96–7*
Davidia involucrata var. *vilmoriniana* 96
deodar 28
Douglas, David 28, 88
dove tree 94, *95*, *96–7*
elm 26
English Landscape Movement 24–5, 26
English oak 30, 70
Eucalyptus chapmaniana 116, *117*
Eucalypytus dalrympleana 116
Evelyn, John 24
Fagus sylvatica 26
Fagus sylvatica 'Purpurea' 30
Farges, Père 94, 114
Flanagan, Mark 118
Flora of China 90, 114
Flowers of the Himalaya (Polunin and Stainton) 109
Frederick, Prince of Wales 26–7, 44
George III, King 27
giant sequoia 28, 86, 88, *87*
Ginkgo biloba *18*, 27, 48, *49*, *50–1*
Gleditsia caspica 82
Global Tree Seed Bank Initiative 30
goat horn tree 114, *115*
Gordon, James (Mile End garden) 25, 27, 44, 48
Gray, Christopher (Fulham garden) 25
Hamilton, Charles (Painshill garden) 24
Handbook on Coniferae (Dallimore and Jackson) 109
handkerchief tree 29, 94, *95*, *96–7*
Hay, Harry 114
Henry, Augustine 94
Holloway Down Nursery (Essex) 70
holm oak 62, 70
Hooker, Joseph Dalton 28
Hooker, Sir William Jackson 28, 76, 82
Indian bean tree 112, *113*
Indian horse chestnut 10, *14*, *19*, *32*, 98, *99*, *100–1*, *102–3*, *104–5*, *106–7*
Italian cypress 22

Jackson, A.B. 108–9
Japanese pagoda tree 7, *18*, 27, 44, *45*, *46–7*
Juglans nigra 30
Juglans regia 23
Kent, William 24–5, *26*
Kirkham, Tony 118
Knight and Perry Nursery (Chelsea) 88
Lambert, A.B. 108
Lancaster, Roy 114
large-leaved alder 82
leaf mining moth 98
Lee and Kennedy Nursery (Hammersmith) 25
lightning, and pine trees 74
Lindley, John 88
Linnaeus, Carl 24, 44, 52
Liquidambar styraciflua 24
Liriodendron chinense 64
Liriodendron tulipfera 19, 24, 64, *65*, *66–7, 68–9*
Lobb, William 86, 88
locust tree 52
London plane 30, 60
London, George 24
Loudon, J.C. 62
Lucombe and Pince Nursery (Exeter) 62
Lucombe oak 28, 62, *63*
Lucombe, William 62
Magnolia 64
Magnolia virginiana 24
maidenhair tree *18*, 27, 48, *49*, *50–1*
Maries, Charles 110
Matthew John D. 86
Maximowicz, Carl Johann 110
Meyer, Carl Anton 82
Millennium Seedbank Partnership 30
mountain gum 116
Natural History (Pliny) 60
Nerium oleander 58
Nesfield, William Andrews 24, 28
Nikko maple 110, *111*
oak *16–17*, 26
oleander 58
oriental plane *15*, *19*, 22, 58, *59*, 60, *61*
Parrotia persica 82
Paulowna, Princess Anna 118
Paulownia kawakamii 8, 118, *119*, *120–1*
Paulownia tomentosa 118
Pearce, Sydney 98
Petre, Lord 24, 25–6
Picea sitchensis 28
Pinus armandi 76, 90, *91*, *92–3*
Pinus bhutanica 108, *109*

Pinus bungeana 76
Pinus koraiensis 76
Pinus nigra 74
Pinus nigra subsp. *laricio* 74, *75*
Pinus nigra subsp. *nigra* 74
Pinus nigra subsp. *pallasiana* 74
Pinus nigra subsp. *salzmannii* 74
Pinus pinea 22, 76, *77*, *78–9*, *80–1*
Pinus wallichiana 90, 108, *109*
Platanus occidentalis 23
Platanus orientalis *15*, 19, 22, 58, *59*, 60, *61*
Platanus × hispanica 30, 60
Pliny the Elder 38, 60
Pococke, Rev. Edward 38, 60
Pococke, Rev. Richard, Bishop of Ossary 38
Psittacula krameri 59
Pterocarya fraxinifolia 82
purple beech 30
Pyrenean pine 74
Quercus castaneifolia 8, *19*, 28, 82, *83*, *84–5*
Quercus castaneifolia 'Green Spire' 82
Quercus cerris 62, 82
Quercus ilex 62, 70
Quercus robur 30, 70, 94, *95*, *96–7*
Quercus spp. 26, *16–17*
Quercus suber 62
Quercus × hispanica 'Lucombeana' 28, 62, *63*
Quercus × turneri 70, *71*, *72–3*
Rehder, Alfred 110
Repton, Humphrey 25
ring-necked parakeet 59
Robin, Jean 52
Robinia pseudoacacia 24, 27, 52, *53*, *54–5*, *56–7*
Royal Botanic Gardens, Kew
    Azalea Garden 64
    Brentford Gate 116
    Broad Walk 28, *29*
    Elizabeth Gate 74
    Great Stove 48
    Ice House 60
    Jodrell Laboratory 116
    Kew Green 98
    Kew Lake 30
    Mediterranean Garden 30
    Orangery 22, 82
    Orangery Lawn 27
    Pagoda *26*
    Pagoda Vista 28
    Palace 58
    Palm House 28, *29*, *31*, 62, 108
    Palm House Pond 60

Pinetum 28, *29*
Princess of Wales Conservatory 44, 70
Rhododendron Dell 30, 60
Ruined Arch 30
Syon Vista 28, 62
Temple of the Sun 22, *23*, 52
Waterlily House 8, *31*, 118
White House 58
Woodland Garden 30
Salisbury, R.A. 74
sapphire dragon tree 8, 118, *119*, *120–1*
Sequoia sempervirens 28, 87, 88
Sequoiadendron giganteum 28, 86, 88, *89*
Siebold, Franz von 118
sitka spruce 28
Sophora japonica 44
southern catalpa 112, *113*
stone pine 22, 76, *77*, *78–9*, *80–1*
Stuart, John: *see* Bute, Lord
Styphnolobium japonicum 7, *18*, 27, 44, *45*, *46–7*
swamp cypress 24
sweet chestnut 26, 34, *35*, *36–7*
Sylva (Evelyn) 24
Taxodium distichum 24
Thorndon (garden) 24, 25–6
Tradescant, John, elder 23–4, 52
Tradescant, John, younger 24, 52
Tree Register of the British Isles 8, 30
tulip tree *19*, 24, 64, *65*, *66–7*, *68–9*
Turkey oak 62, 82
Turner, Spencer 70
Turner, William 22–3, 76
Turner's oak 70, *71*, *72–3*
UK National Tree Seed Project 30
Ulmus 26
umbrella pine 76
Veitch, Sir Harry 94
Veitch's Nursery (Exeter) 88
Veitch's Nursery (Kingston) 110
Veitch's Nursery (Surrey) 86
Wakehurst 30, 116
Wallich, Nathaniel 108, *109*
walnut 23
Walpole, Horace 26
Wellington, first Duke of 88
western catalpa 112
Wharton, Peter 114
Whitton (garden) 24, 27, 52, 58
Wilson, E.H. *29*, 44, 64, 94, 110, 114
Wise, Henry 24
World Heritage Site 7, 22
Yamanaka, Masumi 8, 10, *31*, 98
Zelkova carpinifolia 82

© The Board of Trustees of the Royal Botanic Gardens, Kew 2025

Text © Christina Harrison & Martyn Rix
Illustrations and artworks © Masumi Yamanaka

The authors have asserted their rights to be identified as the authors of this work in accordance with the Copyright, Designs and Patents Act 1988.

All rights reserved. No part of this publication may be reproduced, stored in a retrieval system, or transmitted, in any form, or by any means, electronic, mechanical, photocopying, recording or otherwise, without written permission of the publisher unless in accordance with the provisions of the Copyright Designs and Patents Act 1988.

Great care has been taken to maintain the accuracy of the information contained in this work. However, neither the publisher nor the author can be held responsible for any consequences arising from use of the information contained therein. The views expressed in this work are those of the authors and do not necessarily reflect those of the publisher or of the Board of Trustees of the Royal Botanic Gardens, Kew.

Revised edition published 2025
First published in 2015 by
Royal Botanic Gardens, Kew,
Richmond, Surrey, TW9 3AB, UK

www.kew.org

ISBN 978-1-84246-823-4

Distributed on behalf of the Royal Botanic Gardens, Kew in North America by the University of Chicago Press, 1427 East 60th Street, Chicago, IL 606037, USA.

British Library Cataloguing in Publication Data
A catalogue record for this book is available from the British Library.

Design: Ocky Murray
Original design: Jeff Eden
Production Manager: Georgina Hills
Proofreader: Matthew Seal
Index: Matthew Seal
Printed in Italy by Printer Trento

For information or to purchase all Kew titles please visit shop.kew.org/kewbooksonline or email publishing@kew.org

Kew's mission is to understand and protect plant and fungi, for the wellbeing of people and the future of all life on Earth.

Kew receives approximately one third of its funding from Government through the Department for Environment, Food and Rural Affairs (Defra). All other funding needed to support Kew's vital work comes from members, foundations, donors and commercial activities, including book sales.